ENGINEERING PHYSIOLOGY
Bases of Human Factors/Ergonomics

S E C O N D   E D I T I O N

ENGINEERING PHYSIOLOGY
Bases of Human Factors/Ergonomics

S E C O N D · · · E D I T I O N

K.H.E. KROEMER
H.J. KROEMER
K.E. KROEMER-ELBERT

VAN NOSTRAND REINHOLD
_____ New York

Copyright © 1990 by Van Nostrand Reinhold

Library of Congress Catalog Card Number 90-12238
ISBN 0-442-00354-4

Printed in the United States of America

Van Nostrand Reinhold
115 Fifth Avenue
New York, New York 10003

Van Nostrand International Company Limited
11 New Fetter Lane
London EC4P 4EE, England

Van Nostrand Reinhold
480 La Trobe Street
Melbourne, Victoria 3000, Australia

Nelson Canada
1120 Birchmount Road
Scarborough, Ontario M1K 5G4, Canada

16 15 14 13 12 11 10 9 8 7 6 5 4 3 2 1

Library of Congress Cataloging-in-Publication Data

Kroemer, K. H. E., 1933-
Engineering physiology bases of human factors/ergonomics / K.H.E.
    Kroemer, H.J. Kroemer, K.E. Kroemer-Elbert
        p.   cm.
    Includes bibliographical references.
    ISBN 0-442-00354-4
    1. Human Engineering 2. Anthropometry I. Kroemer, H. J.
(Hiltrud J.) II. Kroemer-Elbert, K .E. (Katrin E.) III. Title.
TA166.K76 1990   90-12238
620.8'2—dc20   CIP

PREFACE

Gunther Lehmann published his *Practical Work Physiology* (in German) in 1952. He used to tell with a smile that his learned colleagues accused him of overly simplifying a difficult subject matter, while many engineers and managers were still baffled by the complexities of the human body. Yet, his book was translated into French, Italian, Spanish, and Polish, and appeared in its second German edition in 1962.

In 1986, we received similar comments on the first edition of our *Engineering Physiology*. The acceptance of the first edition, and its widespread use in courses for engineers and managers have prompted publication of the second edition in 1990.

This second edition is extensively revised. It contains new information on dynamic muscle strength and a chapter on circadian rhythms and work design. But its original purpose is unchanged: to provide physiological information that engineers, designers, and managers need.

We hope to have found a suitable compromise between the necessary depth of information (more can be found in many excellent physiology books) and the desired simplicity and clarity so that the information is "practical."

We thank Ms. Sandy Dalton who expertly and patiently processed this new manuscript.

Radford, VA

Karl H. E. Kroemer
Hiltrud J. Kroemer
Katrin E. Kroemer-Elbert

CONTENTS

CHAPTER 1 ANTHROPOMETRY

CHAPTER 2 THE SKELETAL SYSTEM

CHAPTER 3 SKELETAL MUSCLE

CHAPTER 4 THE NEUROMUSCULAR CONTROL SYSTEM

CHAPTER 5 BIOMECHANICS

CHAPTER 6 THE RESPIRATORY SYSTEM

CHAPTER 7 THE CIRCULATION SYSTEMS

CHAPTER 8 THE METABOLIC SYSTEM

CHAPTER 9 THE THERMAL ENVIRONMENT

Chapter 1

ANTHROPOMETRY

OVERVIEW

Anthropometric* information describes the dimensions of the human body; usually through the use of bony landmarks to which heights, breadths, depths, distances, circumferences, and curvatures are measured. For engineers, the relationship of these dimensions to skeletal "link-joint" systems is of importance, so that the human body can be placed in various positions relative to workstations and equipment. Most available anthropometric information relies on data taken on military and some civilian populations, measured with classical instruments.

The Model

Primary dimensions of the human body are measured to solid identifiable landmarks on bones. Statistical relations among these dimensions, and with others that measure contours, shapes, volumes, and masses help to describe the body.

---

*Anthropometry is derived from the Greek *anthropos* man, and *metrein* to measure.

INTRODUCTION

The dimensions of the human body and its segment proportions have apparently always been of much interest to physicians and anatomists, to rulers and generals, artists and philosophers; to everybody interested in one's neighbor. Marco Polo described his travels around the earth at the end of the 13th century, evoking particular interest with his descriptions of the various body sizes and body builds that he saw in different races and tribes. Physical anthropology as a recording and comparing science is often traced to his travel reports. Around 1800, Johann F. Blumenbach was able to contain the complete anthropometric information available in his book *On the Natural Differences in Mankind.* Alexander von Humboldt encompassed all scientific knowledge present in the five volumes of his *Cosmos* in 1849.

From about the middle of the 19th century, anthropology became diversified into special branches. Adolphe Quételet applied statistics to anthropological information in the middle 1800s. Paul Broca made extensive studies on the skull and drew far-reaching conclusions. Biomechanics was an emerging science at the end of the 19th century — see Chapter 5. The rapidly increasing diversity in anthropometric studies led to conventions of physical anthropologists (1906 in Monaco and 1912 in Geneva) who agreed on standards for anthropometric methods. Rudolf Martin's *Lehrbuch der Anthropologie* (first edition in 1914) quickly became accepted as the authoritative text and handbook for many decades. New engineering needs and developing measuring techniques as well as advanced statistical techniques gave reasons for updating and redirections in the 1960s and 1970s; under H. T. E. Hertzberg and C. E. Clauser, the Anthropology Branch of the U.S. Air Force was one of the driving forces in anthropometric research and engineering applications. Important compilations of anthropometric data, techniques, and methods were published during this period in the U.S.A. (Chapanis 1975; Garrett and Kennedy 1971; Hertzberg 1968; NASA 1978; Roebuck, Kroemer and Thomson 1975; Snyder, Schneider, Owings, Reynolds, Golomb, and Schork 1977).

MEASUREMENT TECHNIQUES

There are four frequently used ways to measure stature: with the subject standing naturally upright, but not stretched; standing freely but stretched to maximum height; leaning against a wall with the back flattened and stretched to maximum height; and lying supine. The differences between the measures when the subject stretches to maximum height either standing freely or leaning are within 2 cm. Lying supine results in a taller measure, but standing "slumped" reduces stature by several centimeters. This example shows that standardization is warranted to assure uniformity in postures and results.

Terminology and Standardization

Body measurements are usually defined by the two end-points of the distance measured, such as elbow-to-fingertip; stature starts at the floor on which the subject stands, and extends to the highest point on the skull.

The following terms are used in anthropometry:

Height is a straight-line, point-to-point vertical measurement.

Breadth is a straight-line, point-to-point horizontal measurement running across the body or a segment.

Depth is a straight-line, point-to-point horizontal measurement running fore-aft the body.

Distance is a straight-line, point-to-point measurement between landmarks on the body.

Curvature is a point-to-point measurement following a contour; this measurement is neither closed nor usually circular.

Circumference is a closed measurement that follows a body contour; hence this measurement usually is not circular.

Reach is a point-to-point measurement following the long axis of the arm or leg.

In general, descriptions of anthropometric measurements include at least three different types of terms. The "locator" identifies the point or landmark on the body whose distance from another point or plane is being measured. The "orientator" identifies the direction of the dimension. The "positioner" designates the body position which the subject assumes for the measurement, such as standing or sitting.

Traditionally, anthropometric measures, such as those previously mentioned, or weight, volume, etc., are taken and reported in the metric system. Professional physical anthropologists are trained to do such measurements; however, with standardization, and training and supervision by an experienced measurer, other non-specialists have successfully taken many measurements. With newer measuring techniques evolving (see later), the demands on both measurers and measuring procedures are changing.

For most measurements, the subject's body is placed in a defined upright straight posture, with body segments at either 180, 0, or 90 degrees to each other. For example, the subject may be required to "stand erect; heels together; buttocks, shoulder blades and back of head touching the wall; arms vertical, fingers straight ...": This is close to the so-called "anatomical position" used in anatomy. The head is positioned in the "Frankfurt Plane": with the pupils on the same horizontal level, the right tragion (approximated by the ear hole) and the lowest point of the right orbit (eye socket) are also placed on the same horizontal plane. When measures are taken on a seated subject, the (flat and horizontal) surfaces of seat and foot support are so arranged that the thighs are horizontal, the lower legs vertical and the feet flat on their horizontal support. The subject is nude, or nearly so, and unshod.

Figure 1-1 shows reference planes and descriptive terms often used in anthropometry. Figure 1-2 illustrates important anatomical landmarks of the human body in the sagittal view, while Figure 1-3 shows landmarks in the frontal view. Figure 1-4 indicates the postures that are typically assumed by the subjects for anthropometric measurements. Publications by the U.S. Department of Defense (1980), Garrett and Kennedy (1971), Hertzberg (1968), NASA (1978), Roebuck, Kroemer, and Thomson (1975), and Wagner (1974; 1988) contain illustrations, descriptions, and definitions of landmarks and measuring techniques.

Classical Measuring Techniques

In conventional anthropometry, the measurement devices are quite simple. The Mourant technique uses primarily a set of grids, often attached to a corner of two orthogonal vertical walls. The subject is placed in front of the grid, and the projections of body landmarks to the grid are used to determine their values — see Figure 1-5. Other setups may include box-like jigs which provide references for measurements of head and foot dimensions — see Figures 1-6 and 1-8. Many body landmarks, however, cannot easily be projected onto grids. Here, special instruments are available, shown in Figure 1-7. The most important is the anthropometer, which consists basically of a graduated

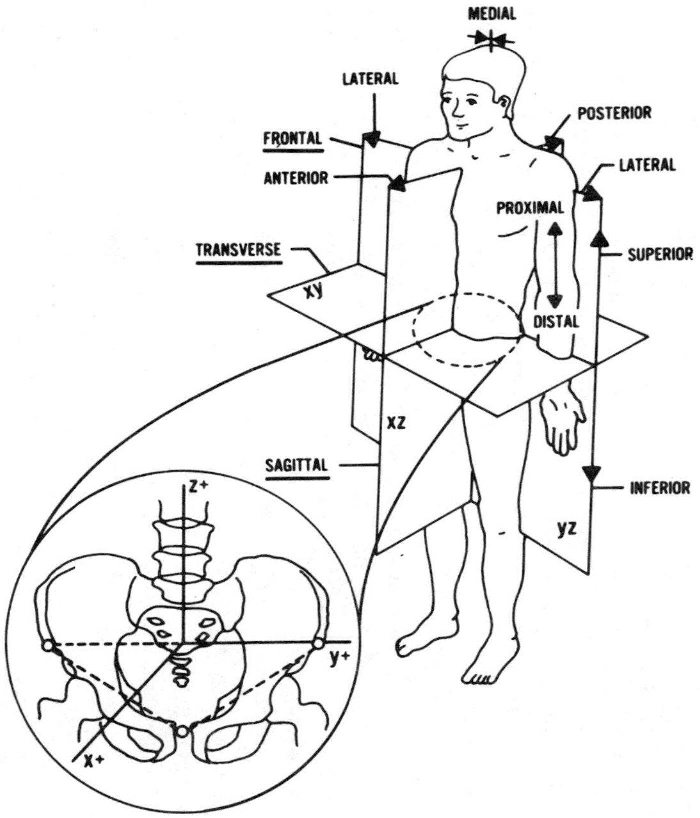

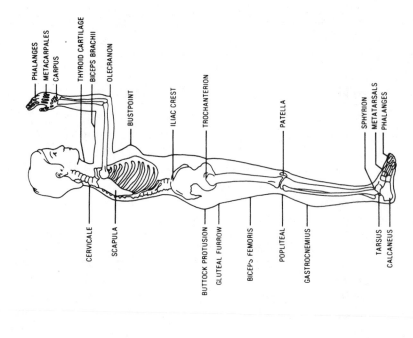

Figure 1-2. Anatomical landmarks in the sagittal view (NASA 1978).

Figure 1-1. Descriptive terms and measuring planes used in anthropometry (NASA 1978).

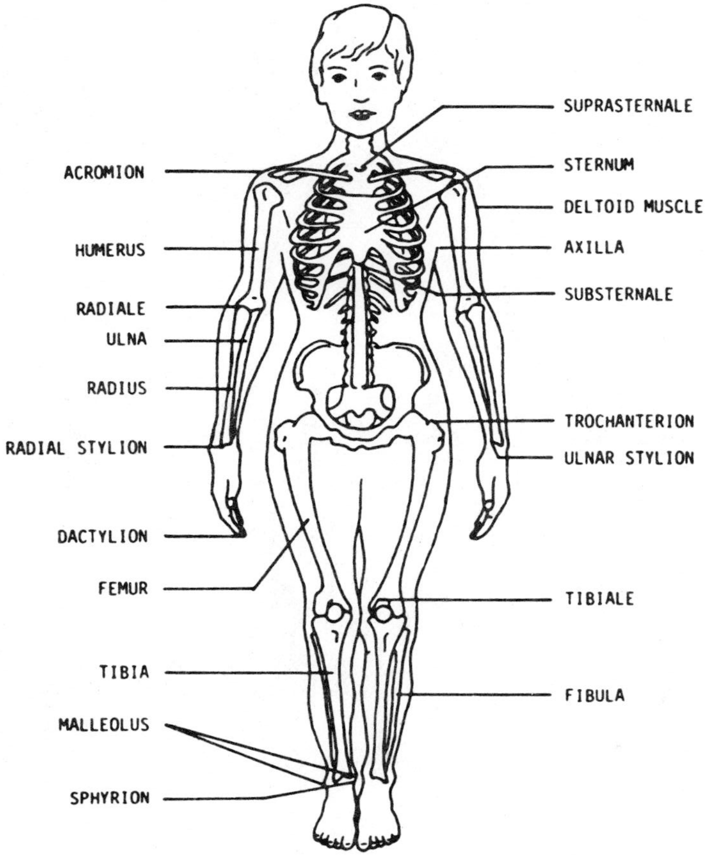

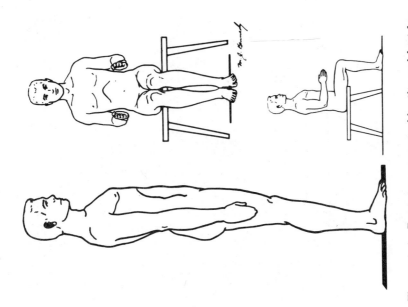

Figure 1-3. Anatomical landmarks in the frontal view (NASA 1978).

Figure 1-4. Postures assumed by the subject for anthropometric measurements.

SUPRASTERNALE
STERNUM
DELTOID MUSCLE
AXILLA
SUBSTERNALE
TROCHANTERION
ULNAR STYLION
TIBIALE
FIBULA

ACROMION
HUMERUS
RADIALE
ULNA
RADIUS
RADIAL STYLION
DACTYLION
FEMUR
TIBIA
MALLEOLUS
SPHYRION

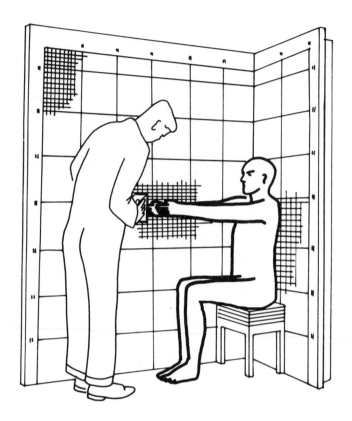

Figure 1-5. Grid system placed in a corner for anthropometric measurements (adapted from Roebuck, Kroemer, and Thomson 1975).

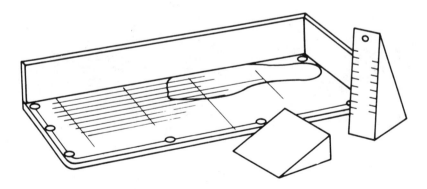

Figure 1-6. "Measuring box" for foot measurements (adapted from Roebuck, Kroemer, and Thomson 1975).

rod with a sliding branch at right angle to the rod. This sliding branch allows to reach behind corners and folds of the body whose distances from the reference are read from the scaled rod. The rod itself can usually be sectioned for ease of transport and storage. The anthropometer has one branch permanently affixed to its top part. This allows distance measurements between the fixed and the sliding branches: this setup is called a sliding caliper, or compass. The spreading caliper consist of two curved branches which are joined in a hinge. The distance between the tips of the two opposing ends are read on a scale which is attached in a convenient place between the two parts of the caliper. Many other devices are also employed: a small sliding caliper serves for short measurements, e.g., of finger thicknesses or finger lengths; a special caliper measures the thickness of skinfolds; a cone is used to measure the diameter around which fingers can close; circular holes of increasing sizes drilled in a thin plate indicate external finger diameters. Circumferences and curvatures are measured with tapes, usually made of flat steel, occasionally of nonstretching woven material. Figures 1-8 to 1-10 show typical uses of anthropometer, caliper, and tape. Scales, of course, are used to measure the weight of the whole body, or of body segments. Many other measuring methods have been applied in special cases, such as the shadow technique, templates, multiple probes, casting, etc. They are explained by Roebuck, Kroemer, and Thomson (1975).

Emerging Measurement Methods

The classical anthropometric techniques rely on measurement instruments applied by the hand of the measurer to the body of the subject; they are simple, relatively easily used, and inexpensive. However, their use is somewhat clumsy and certainly time-consuming; each measurement and tool must be selected in advance; and what was not measured in the test session remains unknown unless the subject is called back for more measuring. A major disadvantage is also that many of the classical dimensions are not related to each other in space. For example, as one looks at a subject from the side, stature, eye height, and shoulder height are located in different yet undefined frontal planes. Furthermore, certain parts of the body are very sensitive and cannot be touched, such as the eyes.

Photographic methods overcome many of these disadvantages. In particular, they can record all three-dimensional aspects of the human body. They allow storage of practically infinite numbers of measurements which can be taken at one's convenience from the record. However, they also have drawbacks: the equipment (particularly for data analyses) is expensive: a scale may be difficult to establish; parallax distortions must be eliminated; landmarks cannot be palpated under the skin on the photograph.

It is for these and other reasons that until the 1970s photographic anthropometry was not commonly used, although systematic attempts had been made at least a century earlier. Recently, many improvements have been introduced by use of several cameras, of mirrors, of film and videotape instead of still photography, by holography, and various methods of stereophotometry. Based on the pioneering efforts of Herron (1973) and other proponents, most of these methods extract a large number of data points from the recorded picture of the body and integrate these, using computer programs, to entire models of the human body. These then allow the extraction of desired dimensions and contour identifiers (see, e.g., Baughman 1982 or Drerup and Hierholzer 1985).

Other techniques (often used in biomechanical analyses as well) rely on the identification of certain points of interest, such as body articulations, by "markers" which are

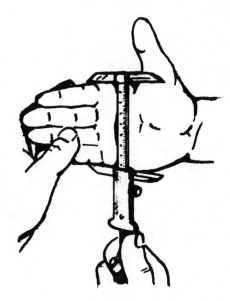

Figure 1-9. Measuring with a sliding caliper, and using the anthropometer as a sliding rule.

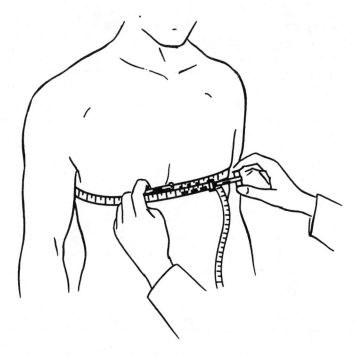

Figure 1-10. Circumference measurement with a tape.

traced on the record (photographic, videotape, etc.) made of the body in motion. The markers can simply provide an optical contract to the background, or they may actively emit light or sonic signals. Obviously, the technology in this field is very much in flux. Laser techniques are promising. Classical measuring techniques are likely to retain some importance because they are so simple and easily applied even under adverse conditions, particularly in static or semi-static situations. However, high-technology anthropometric methods will gain increasing importance quickly.

Body Typology

One can try to assess human body dimensions by using descriptors of body components, and how they "fit" together; our images of the beautiful body are affected by aesthetic codes, canons and rules founded on often ancient (e.g., Egyptian, Greek, Roman) concepts of the human body. A more recent example is Leonardo de Vinci's drawing of the body within a frame of graduated circles and squares as a symbol of the well-proportioned body; it has been adopted, simplified, as the emblem of the U.S. Human Factors Society.

Categorization of body builds into different types is called somatotyping, from the Greek *soma* for body. Hippocrates developed, about 400 B.C., a scheme that included four body types which were supposedly determined by their fluids. (The "moist" type was dominated by "black gall"; the "dry" type was governed by "yellow gall"; the "cold" type was characterized by its "slime", while the "warm" type was governed by blood.) In 1921 the psychiatrist Ernst Kretschmer published a system of three body types intended to relate body build to personality traits ("Koerperbau and Charakter"). Kretschmer's typology consisted of the asthenic, pyknic and the athletic body build. (The "athletic" type was to indicate character traits, not sportive performance capabilities.) In the 1940s, the anthropologist W. H. Sheldon established a system of three body types which indeed was intended to describe (male) body proportions. Sheldon rated each person's appearance with respect to ecto-, endo- and mesomorphic components — see Table 1-1. It is of interest to note that Sheldon's typology was originally based on intuitive assessment, not on actual body measurements, which were introduced into the system by his disciples.

Unfortunately, these and other attempts at somatotyping (such as the Heath—Carter system) have not provided reliable predictors of attitudes nor of capabilities or limitations regarding human performance in technological systems. Hence, somatotyping is of little value for engineers or managers.

Table 1-1. Body typologies.

Descriptors	Stocky, Stout, Soft, Round	Strong, Muscular, Sturdy	Lean, Slender, Fragil
Kretschmer typology	pyknic	athletic	asthenic (leptosomic)
Sheldon typology	endomorphic	mesomorphic	ectomorphic

ANTHROPOMETRIC DATA

The military has always had a particular interest in the body dimensions of soldiers — for a variety of reasons, among them the necessity to provide fitting uniforms, armor, and equipment. Furthermore, armies had and have their surgeons and medical personnel willing and capable to perform body measurements on large samples "on command." Hence, anthropometric information about soldiers has a long history and is rather complete: the anthropometric data bank of the U.S. Air Force Aerospace Medical Research Laboratory (Wright-Patterson AFB, OH 45433) contains the data of approximately 100 military or civilian surveys from many nations, but the majority of the data consists of dimensions measured on U.S. Navy, Army, and Air Force personnel. The NASA Anthropometric Sourcebook (1978) provides a good overview of the data available up to the mid-1970s. In 1980 the Department of Defense published a compendium of U.S. military anthropometry, while Kennedy collated in 1986 the data available on U.S. Air Force personnel. A report on U.S. Army personnel was compiled in 1989 (Gordon, Churchill, Clauser, Bradtmiller, McConville, Tebbetts, and Walker). Dimensions of the head, hand, and foot measured on military populations are apparently similar to those of civilians (NASA 1975; Garrett 1971; White 1982) while other dimensions reflect that soldiers are a select sample of the general population.

Practical Statistics

Normality

Anthropometric data are, fortunately, usually distributed in a reasonably normal (Gaussian) distribution. Hence, regular parametric statistics apply in most cases. The most common statistical procedures are listed in Table 1-2.

The first and easiest check is on mean, median, and mode: if they concur, normality is likely. Another method is to calculate the average by first using the complete range of data and then by leaving out the (say, ten) smallest and largest numbers: if both calculations end up with the same mean, normality is often the case. More formal calculations rely on the measures of symmetry and of peakedness contained in Table 1-2. Regarding skewness: a result of 0 indicates a symmetrical distribution, a positive (negative) result points to skewness to the left (right). Regarding peakedness (kurtosis), a normal distribution is usually assumed when the numerical result is 3; if the result is larger (smaller) than 3 the distribution is peaked (flat).

Variability

The standard deviation is a very useful statistic for many purposes. It describes the dispersion of a given set of anthropometric data, which reflects the (true) variability of the underlying data and the accuracy and reliability of the measuring technique applied. For example: if one compares the descriptive statistics of three distinct though related population samples, and finds that one shows a much larger coefficient of variation (standard deviation divided by the mean) of like dimensions than the others, one deduces from this result that either the sample with the large dispersion was in fact more variable than the other two samples, or that there was suspicious variability in the measuring technique or in the data recording.

Table 1-2. Statistical formulas of particular use in anthropometry.

MEASURES OF CENTRAL TENDENCY

Mean, Average	$\bar{x} = \Sigma x/N$	(1st Moment)	(A-1)
Median	Middle value (of values in numerical order), 50th percentile.		
Mode	Most often found value.		

MEASURES OF VARIABILITY

Range	$x_{max} - x_{min}$		(A-2)
Standard Deviation	$S = (\text{Variance})^{1/2} = [\Sigma(x - \bar{x})^2/(N - 1)]^{1/2}$	(2nd Moment)	(A-3)
Coefficient of Variation	$CV = S/\bar{x}$		(A-4)
Standard Error of the Mean	$SE = S/N^{1/2}$		(A-5)
Two-Sided Confidence Limits For Mean	$\bar{x} \pm k\ SE$ (see Table A-3 for k)		(A-6)
Skewness	$\Sigma(x - \bar{x})^3/N$	(3rd Moment)	(A-7)
Peakedness	$\Sigma(x - \bar{x})^4/N$	(4th Moment)	(A-8)

MEASURES OF RELATIONSHIP BETWEEN TWO VARIABLES X AND Y

Correlation Coefficient	$r = S_{xy}/(S_x S_y)^{1/2} =$ $\Sigma[(x - \bar{x})(y - \bar{y})]/[\Sigma(x - \bar{x})^2 \Sigma(y - \bar{y})^2]^{1/2}$	(A-9)
Regression	$y = ax + b$	(A-10)
	$a = r\ S_y/S_x$	(A-11)
	$SE_y = S_y(1 - r^2)^{1/2}$	(A-12)

Percentiles

Of the other statistical measures listed in Table 1-2, some are of extraordinary importance and usefulness. Of course, the mean (average) value allows one to compute with help of the standard deviation any desired percentile value along the (normal) distribution of the variable. Table 1-3 lists multipliers and the percentile values at which one arrives using those k factors. For example: to determine the 5th percentile value, one selects from the table $k = 1.65$ by which one multiplies the standard deviation; this product, subtracted from the mean, gives the value of the variable x at the 5th percentile (x_{p5}).

Percentiles are very convenient to determine exactly which percentage of a known population a given measure x identifies; likewise, if two percentile values along the continuum of one variable have been selected, one knows exactly which percentile range of the population will be fitted between these extreme values (e.g., x_{p5} and x_{p95}). If one intends, for example, to fit persons between the 5th and 95th percentile, it is obvious that this design will be too large for the smallest five percent and too small for the largest five percent; the selected range x_{p5} to x_{p95} accommodates exactly the central 90 percent. The use of percentiles is a significant step away from the attractively simple but nevertheless false assumption that one should or could design for the mythical "average person": there is no person of all 50th percentile body dimensions. (Of course, the all 5th percentile "little woman" or all 95th percentile "big guy" do not exist either.)

Correlations

Some body dimensions are closely related with each other: for example, eye height is very highly correlated with stature — but head length is not, nor waist circumference. Table 1-4 shows selected correlation coefficients among body dimensions of U.S. Air Force personnel, male and female. (A more detailed table is contained in the Appendix to this chapter.)

Table 1-3. Calculation of percentiles.

Percentile p associated with			
$x_i = \bar{x} - kS$ (below mean)	$x_j = \bar{x} + kS$ (above mean)	Central Percent Included in the range x_{pi} to x_{pj}	k
0.5	99.5	99	2.576
1	99	98	2.326
2	98	96	2.06
2.5	97.5	95	1.96
3	97	94	1.88
5	95	90	1.65
10	90	80	1.28
15	85	70	1.04
16.5	83.5	67	1.00
20	80	60	0.84
25	75	50	0.67
37.5	62.5	25	0.32
50	50	0	0

Table 1-4. Selected correlation coefficients for anthropometric data on U.S. Air Force personnel: women above the diagonal, men below (from NASA 1978).

		1	2	3	4	5	6	7	8	9	10
1.	Age		.223	.048	-.023	.039	-.055	.091	-.072	.233	.287
2.	Weight	.113		.533	.457	.497	.431	.481	.370	.835	.799
3.	Stature	-.028	.515		.927	.914	.849	.801	.728	.334	.257
4.	Chest height	-.028	.483	.949		.897	.862	.673	.731	.271	.183
5.	Waist height	-.033	.422	.923	.930		.909	.607	.762	.308	.238
6.	Crotch height	-.093	.359	.856	.866	.905		.467	.788	.264	.190
7.	Sitting height	-.054	.457	.786	.681	.580	.453		.398	.312	.239
8.	Popliteal height	-.102	.299	.841	.843	.883	.880	.485		.230	.172
9.	Shoulder circumference	.091	.831	.318	.300	.261	.212	.291	.182		.810
10.	Chest circumference	.259	.832	.240	.245	.203	.147	.171	.114	.822	
11.	Waist circumference	.262	.856	.224	.212	.142	.132	.167	.068	.720	.804
12.	Buttock circumference	.105	.922	.362	.334	.278	.217	.347	.149	.744	.766
13.	Biacromial breadth	.003	.452	.378	.335	.339	.282	.349	.316	.555	.401
14.	Waist breadth	.214	.852	.287	.260	.215	.195	.216	.133	.715	.801
15.	Hip breadth	.105	.809	.414	.380	.342	.283	.376	.221	.632	.647
16.	Head circumference	.110	.412	.294	.251	.233	.188	.287	.194	.327	.340
17.	Head length	.054	.261	.249	.218	.208	.170	.244	.175	.201	.196
18.	Head breadth	.122	.305	.133	.097	.089	.066	.132	.075	.245	.271
19.	Face length	.119	.228	.275	.220	.226	.199	.253	.193	.162	.172
20.	Face breadth	.233	.453	.190	.160	.142	.099	.185	.098	.401	.421

		11	12	13	14	15	16	17	18	19	20
1.	Age	.234	.219	.149	.146	.194	.095	.118	.190	.189	.089
2.	Weight	.824	.886	.495	.768	.770	.403	.304	.290	.264	.358
3.	Stature	.279	.360	.456	.329	.348	.331	.318	.136	.267	.199
4.	Chest height	.216	.289	.412	.266	.276	.284	.284	.085	.222	.162
5.	Waist height	.238	.336	.409	.293	.318	.306	.297	.123	.225	.200
6.	Crotch height	.221	.246	.380	.277	.225	.294	.280	.089	.205	.172
7.	Sitting height	.236	.383	.384	.277	.379	.294	.275	.136	.248	.146
8.	Popliteal height	.186	.201	.327	.249	.181	.235	.253	.087	.185	.189
9.	Shoulder circumference	.775	.717	.581	.719	.606	.330	.248	.252	.217	.313
10.	Chest circumference	.796	.674	.370	.706	.551	.273	.204	.255	.176	.273
11.	Waist circumference		.722	.382	.886	.600	.281	.149	.267	.174	.310
12.	Buttock circumference	.852		.396	.668	.893	.310	.214	.238	.180	.269
13.	Biacromial breadth	.288	.355		.401	.361	.311	.239	.178	.266	.211
14.	Waist breadth	.936	.849	.327		.576	.292	.168	.263	.182	.296
15.	Hip breadth	.724	.895	.340	.760		.265	.183	.188	.155	.215
16.	Head circumference	.309	.330	.251	.310	.288		.692	.430	.273	.299
17.	Head length	.158	.195	.179	.164	.166	.779		.115	.311	.113
18.	Head breadth	.265	.252	.188	.268	.227	.521	.058		.174	.497
19.	Face length	.129	.186	.187	.151	.161	.315	.289	.148		.144
20.	Face breadth	.412	.394	.278	.410	.364	.464	.131	.660	.206	

The correlation coefficient provides concise information about the relationship between two or more sets of data. It is common practice in anthropometry (in fact in human engineering altogether) to require a coefficient of at least 0.7 in order to base design decisions on this correlation. The reason for selecting this "coefficient of determination" as the minimum acceptable value lies in the convention that one should be able to explain at least 50% of the variance of the predicted value from the predictor variable: this requires r^2 to be at least 0.5, i.e., r at least 0.7075. (Note that r depends on sample size N.)

Regressions

This "0.7 convention" is important for development and use of regression equations: they express the average of one variable as a function of another — see Table 1-2. If the use of only one predictor variable is not sufficient to establish an overall correlation coefficient between predictor and predicted value of at least 0.7, additional predictor variables may be taken into the equation until that minimal cut-off point is exceeded. (Examples for such regression equations are in Table 1-7.)

Statistical Body Models

It is often desired to combine body dimensions to construct a complete model of the human body, or of its major components. For this purpose, two methods have been employed: one is incorrect, but the other is suitable.

The percentile statistic is convenient for establishing the location of one given datum measured along its continuum range, e.g., 149.5 cm as the 5th percentile value for female stature. However, it is *false* to believe that all other body component measures of that person must also be at their 5th percentile; p5 leg length plus p5 torso length plus p5 head height do *not* add up to p5 stature (Robinette and McConville 1981): a person of p5 stature may have relatively short legs but a long torso or vice versa.

Regression equations, however, are suitable to generate discrete body measures, using as predictor variables other dimensions whose values are known for the population sample of interest. Given that the equations are proper, predicted values can be added or subtracted. For example, 5th percentile values for leg, trunk, and head heights predicted from regressions do add up to the correct p5 stature.

Hence, percentiles are useful descriptors of discrete body measures, but they can be "stacked" only if derived from regression equations concerning the sample in question.

Sample Size

If one needs to take measurements on a sample, one usually tries — for a variety of reasons — to keep the sample at the smallest acceptable number. Assuming normal distribution of the measured variable, the required sample size N can be estimated from

$$N = S^2 m^2 d^2 \tag{1-1}$$

where S is the (estimated) standard deviation of the data, d the desired accuracy (in ± d units) of the measurement, and m is taken from Table 1-5. (Note that the multiplier values m depend on the statistic of interest.)

If the initially calculated sample size N is below 100, the values for m given in Table 1-5 should be replaced by

$m = 2.00$ for $100 > N > 40$
$m = 2.05$ for $40 > N > 20$
$m = 2.16$ for $20 > N > 10$
$m = 2.78$ for $10 > N$

and the calculation be repeated (Roebuck, Kroemer, and Thomson 1975).

Estimating the Standard Deviation

Often it is necessary or desirable to estimate the standard deviation of a population statistic. One can make reasonable estimates in several ways, some of which are based on statistical procedures, others are more empirically based on known biometric patterns.

One may use the standard deviation of the same dimension found in other populations to predict the sample value. This obviously assumes that the dimension in question has about the same distribution in similar populations. Of particular value for this is the measure of relative dispersion, that is the coefficient of variation in Table 1-2.

If one needs to know the coefficient of variation of either the sum or the difference between two dimensions with known dispersion, and if the correlation coefficient between the dimensions is known, the following equations can be used (normal distribution assumed):

$$S_{(x+y)} = (S_x^2 + S_y^2 + 2r\,S_x S_y)^{1/2} \qquad (1\text{-}2)$$

For subtraction of two dimensions

$$S_{(x-y)} = (S_x^2 + S_y^2 - 2r\,S_x S_y)^{1/2} \qquad (1\text{-}3)$$

Table 1-5. Values of m for sample size determination.

m	Statistic of Interest
1.96	Mean
1.39	Standard deviation
2.46	50th percentile
2.46	45th and 55th percentile
2.49	40th and 60th percentile
2.52	35th and 65th percentile
2.58	30th and 70th percentile
2.67	25th and 75th percentile
2.80	20th and 80th percentile
3.00	15th and 85th percentile
3.35	10th and 90th percentile
4.14	5th and 95th percentile
4.46	4th and 96th percentile
4.92	3rd and 97th percentile
5.67	2nd and 98th percentile
7.33	1st and 99th percentile

The short Table 1-4 or the extensive correlation table in the Appendix to this chapter may be of help in estimating variability.

Composite Population

It may be necessary to consider a population that consists of two distinct and known subsamples: an example is to design for a user group that consists of $a\%$ females and $b\%$ males, with $a + b = 100\%$. To determine at what percentile of the composite population a specific value of x is, one proceeds stepwise as follows (Kroemer 1983):

Step 1: Determine k factors associated with x in the samples a and b.

For sample a:

$$x_a = \bar{x}_a - k_a S_a \quad \text{if } x_a < \bar{x}_a$$

$$x_a = \bar{x}_a + k_a S_a \quad \text{if } x_a > \bar{x}_a \tag{1-4}$$

$$k_a = |x_a - \bar{x}_a| \, S_a^{-1}$$

Similarly, for sample b:

$$k_b = |x_b - \bar{x}_b| \, S_b^{-1} \tag{1-5}$$

Step 2: Obtain factor k associated with x in the combined population:

$$k = ak_a + bk_b \tag{1-6}$$

Step 3: Determine percentile p associated with k: use Table 1-3.

If percentiles p for each x are known in each group, one may simply add the proportioned percentiles:

$$p = ap_a + bp_b \tag{1-7}$$

Body Proportions

In the past, some people have found it convenient to calculate ratios or proportional relationships of body dimensions: these are often called indices. Such index numbers are unjustified and misleading when there is an insufficient correlation between the two variables — see above. About three decades ago Drillis and Contini published a "stickman" figure that indicated ratios between various body segments and stature; reprinted (with appropriate warnings and notes of caution) by Roebuck, Kroemer, and Thomson (1975) and by Chaffin and Andersson (1984). Unfortunately, some practitioners have used these ratios indiscriminately even when the coefficients of correlation were well below the "0.7 convention." In Table 1-4 only chest height, waist height, crotch height (a measure of leg length), sitting height, and popliteal height show correlation coefficients of more than 0.7 to stature; head length, shoulder breadth, hip breadth, and many other variables are not well correlated with stature. Hence, wanting to predict one dimension from another, one must carefully check whether the correlation between the two data sets is sufficiently high. In general, long bone (link) dimensions are reasonably well correlated with each other. Breadth and depth dimensions, respectively, correlate reasonably well within their groups, but not with stature; the same is true for circumferences and for foot, hand, and head measures — see the correlation Table 1-10 in the Appendix to this chapter for more data.

Anthropometric Data of U.S. Civilians

Measured anthropometric data on civilian populations are rather sparse while the body dimensions of military personnel are well known. Since the military is a large albeit biased (young and healthy) sample of the overall population, it appears logical to use the military data to infer dimensions for the general civilian population. There are, however, several problems involved: One is that military personnel may be so highly selected that they constitute a special sample which is not representative of the overall population. Another question is whether a sufficient number of dimensions was measured both in the military and the general population to allow an assessment of whether or not one data set can represent the other. McConville, Robinette, and Churchill (1981) addressed both problems. They selected the U.S. Health Examination Survey (HES 1965) and compared it for males to the 1967 survey of the U.S. Air Force and the U.S. Army 1966 survey; the comparison data for females came from the 1968 U.S. Air Force and the 1977 U.S. Army surveys. The underlying assumption for the comparisons was that if good height and weight matches can be achieved between civilian and military individuals, then the means and standard deviations of other dimensions (measured in either survey) should be well matched also. (Of course, this is a debatable assumption.)

The procedure used was to match individuals from the civilian and military surveys on the basis of stature and weight (with matching intervals of ±1 inch and ±5 pounds). Thus, a new military sample was created which represented the civilian sample in height and weight. From the "new" (matched) military sample, dimensions other than height and weight were selected and compared to the corresponding data measured in the civilian surveys.

For the males, an excellent fit was achieved: ninety-nine percent of all civilian individuals could be matched with at least one soldier. The mean differences of the samples in stature and weight were negligible. A comparison of six linear dimensions measures both in the military and civilian surveys provided similar good matches in means and standard deviations.

Use of regression equations to calculate one body dimension from others (height and weight) resulted in averages of the six civilian dimensions predicted from the military set which were very close to the results achieved in matching. However, standard deviations predicted from regression equations turned out to be considerably larger than those obtained in the matched pair procedure.

The same procedures were used in comparing female civilian and military data. This was almost as successful as for the male data. About 94% of the civilian data could be matched with military individuals, with good fit in stature but poorer matches in weight. (The unmatched individuals were very short and very heavy civilians with weights between 90 and 135 kg. For these persons, no military matches were found.) Hence, with proper caution and insight one can use military anthropometric data to determine data of the general population.

The anthropometric information on the U.S. civilian population in Table 1-6 was compiled in 1980 from measured, matched, and predicted data by Drs. Kennedy and McConville for publication by Kroemer in 1981. Table 1-7 lists the major regression equations used to calculate many of the data. Table 1-6 should well reflect the dimensions in the 1960's and 1970's; stature has probably been increasing since by about 1 cm

Table 1-6. U.S. civilian body dimensions, female/male, in cm for ages 20 to 60 years.*

	5th	50th	95th	Std. Dev.
		Percentiles		
HEIGHTS(f above floor, s above seat)				
Stature (Height)^f	149.5 / 161.8	160.5 / 173.6	171.3 / 184.4	6.6 / 6.9
Eye Height^f	138.3 / 151.1	148.9 / 162.4	159.3 / 172.7	6.4 / 6.6**
Shoulder (acromion) Height^f	121.1 / 132.3	131.1 / 142.8	141.9 / 152.4	6.1 / 6.1**
Elbow Height^f	93.6 / 100.0	101.2 / 109.9	108.8 / 119.0	4.6 / 5.8
Knuckle Height^f	64.3 / 69.8	70.2 / 75.4	75.9 / 80.4	3.5 / 3.2
Height, sitting^s	78.6 / 84.2	85.0 / 90.6	90.7 / 96.7	3.5 / 3.7
Eye Height, sitting^s	67.5 / 72.6	73.3 / 78.6	78.5 / 84.4	3.3 / 3.6**
Shoulder Height, sitting^s	49.2 / 52.7	55.7 / 59.4	61.7 / 65.8	3.8 / 4.0**
Elbow Rest Height, sitting^s	18.1 / 19.0	23.3 / 24.3	28.1 / 29.4	2.9 / 3.0
Knee Height, sitting^f	45.2 / 49.3	49.8 / 54.3	54.5 / 59.3	2.7 / 2.9
Popliteal Height, sitting^f	35.5 / 39.2	39.8 / 44.2	44.3 / 48.8	2.6 / 2.8
Thigh Clearance Height^f	10.6 / 11.4	13.7 / 14.4	17.5 / 17.7	1.8 / 1.7
DEPTHS				
Chest Depth	21.4 / 21.4	24.2 / 24.2	29.7 / 27.6	2.5 / 1.9**
Elbow-Fingertip Distance	38.5 / 44.1	42.1 / 47.9	56.0 / 51.4	2.2 / 2.2
Buttock-Knee Distance, sitting	51.8 / 54.0	56.9 / 59.4	62.5 / 64.2	3.1 / 3.0
Buttock-Popliteal Distance, sitting	43.0 / 44.2	48.1 / 49.5	53.5 / 54.8	3.1 / 3.0
Forward Reach, Functional	64.0 / 76.3	71.0 / 82.5	79.0 / 88.3	4.5 / 5.0
BREADTHS				
Elbow-to-Elbow Breadth	31.5 / 35.0	38.4 / 41.7	49.1 / 50.6	5.4 / 4.6
Hip Breadth, sitting	31.2 / 30.8	36.4 / 35.4	43.7 / 40.6	3.7 / 2.8
HEAD DIMENSIONS				
Head Breadth	13.6 / 14.4	14.54 / 15.42	15.5 / 16.4	.57 / .59
Head Circumference	52.3 / 53.8	54.9 / 56.8	57.7 / 59.3	1.63 / 1.68
Interpupillary Distance	5.1 / 5.5	5.83 / 6.20	6.5 / 6.8	.44 / .39
FOOT DIMENSIONS				
Foot Length	22.3 / 24.8	24.1 / 26.9	26.2 / 29.0	1.19 / 1.28
Foot Breadth	8.1 / 9.0	8.84 / 9.79	9.7 / 10.7	.50 / .53
Lateral Malleolus Height	5.8 / 6.2	6.78 / 7.03	7.8 / 8.0	.59 / .54
HAND DIMENSIONS				
Hand Length	16.4 / 17.6	17.95 / 19.05	19.8 / 20.6	1.04 / .93
Breadth, Metacarpale	7.0 / 8.2	7.66 / 8.88	8.4 / 9.8	.41 / .47
Circumference, Metacarpale	16.9 / 19.9	18.36 / 21.55	19.9 / 23.5	.89 / 1.09
Thickness, Meta III	2.5 / 2.4	2.77 / 2.76	3.1 / 3.1	.18 / .21
Digit 1: Breadth,				
Interphalangeal	1.7 / 2.1	1.98 / 2.29	2.1 / 2.5	.12 / .21
Crotch-Tip Length	4.7 / 5.1	5.36 / 5.88	6.1 / 6.6	.44 / .45
Digit 2: Breadth,				
Distal Joint	1.4 / 1.7	1.55 / 1.85	1.7 / 2.0	.10 / .12
Crotch-Tip Length	6.1 / 6.8	6.88 / 7.52	7.8 / 8.2	.52 / .46
Digit 3: Breadth,				
Distal Joint	1.4 / 1.7	1.53 / 1.85	1.7 / 2.0	.09 / .12
Crotch-Tip Length	7.0 / 7.8	7.77 / 8.53	8.7 / 9.5	.51 / .51
Digit 4: Breadth,				
Distal Joint	1.3 / 1.6	1.42 / 1.70	1.6 / 1.9	.09 / .11
Crotch-Tip Length	6.5 / 7.4	7.29 / 7.99	8.2 / 8.9	.53 / .47
Digit 5: Breadth,				
Distal Joint	1.2 / 1.4	1.32 / 1.57	1.5 / 1.8	.09 / .12
Crotch-Tip Length	4.8 / 5.4	5.44 / 6.08	6.2 / 6.99	.44 / .47
WEIGHT (in kg)	46.2 / 56.2	61.1 / 74.0	89.9 / 97.1	13.8 / 12.6

*   Courtesy of Dr. J. T. McConville, Anthropology Research Project, Yellow Springs OH 45387 and Dr. K. W. Kennedy, then USAF-AAMRL-HEG, OH 45433.

**   Estimated by K. H. E. Kroemer.

Table 1-7. Regression equations *(all data in cm, except weight in pounds).

FOR MALES:

Variable Predicted	Equations	Std. Error of Estimate	Resulting Correlation
Eye Height, standing	= (.9544 Stature) – 32.9	0.99	0.986
Acromion Height	= (.927 Stature) – (.233 Sitting Height) + (.042 Chest Circumference) – 8.1318	1.79	0.957
Elbow Height	= (.879 Stature) – (.629 Sitting Height) + (.674 Elbow Rest Height sitting) – 2.0578	1.29	0.960
Knuckle Height	= (.5536 Stature) – (.1982 Knee Height sitting) – 9.961	2.08	0.820
Hip Breadth	= (.688 Weight) + (.88 Sitting Height) – .0062	1.54	0.878
Functional Reach Forward	= (.102 Stature) + (.497 Knee Height sitting) + (.461 Buttock-Knee Length) + 10.4423	3.99	0.613
Vertical Reach Sitting	= (.8883 Stature) + (.2525 Sitting Height) – 39.70	3.13	0.842
Eye Height, sitting	= (.827 Sitting Height) + (.79 Elbow Rest Height sitting) + (.60 Buttock-Knee Length) – 1.769	1.09	0.933
Acromion Height, sitting	= (.202 Stature) + (.93 Sitting Height) + (.709 Elbow Rest Height sitting) – 1.2814	1.01	0.936
Forearm-Hand Length	= (.086 Stature) + (.298 Knee Height sitting) + (.234 Buttock-Knee Length) + 2.8683	1.32	0.821
Chest Depth	= (.263 Weight) – (.049 Sitting Height) + (.165 Chest Circumference) + 7.9929	1.17	0.811

FOR FEMALES:

Variable Predicted	Equations	Std. Error of Estimate	Resulting Correlation
Eye Height, standing	= (.963 Stature) – 5.7101	1.07	0.984
Acromion Height	= (.957 Stature) – (.208 Sitting Height) + (.065 waist circumference) – 9.6449	1.47	0.967
Elbow Height	= (.6952 Stature) – 10.33	1.75	0.931
Knuckle Height	= (.4095 Stature) + (.2227 Sitting Height) – 14.371	2.19	0.833
Hip Breadth	= (1.338 Weight) – (.148 Popliteal Height) – (.100 Chest Circumerence Scye) + 32.4839	1.39	0.801
Overhead Reach, standing	= (1.066 Stature) – (.287 Elbow Rest Height) + (.505 Buttock-Knee Length) + 3.1681	4.28	0.866
Eye Height, sitting	= (.907 Sitting Height) – 3.7877	1.07	0.943
Elbow-Fingertip Length	= (.643 Popliteal Height) + (.175 Buttock-Knee Length) + 6.5701	1.19	0.855
Chest Circum. at Nipple	= (.8381 Chest Circumference at Scye) + (.0861 Weight) + 5.727	2.85	0.896
Chest Depth	= (.067 Stature) + (.613 Weight) + (.147 Chest Circumference at Scye) + 13.9110	1.34	0.766

*Modified from K. M. Robinette's personal communication of 10/22/81.

per decade — see the discussion of secular change. Nevertheless, the data in this table are currently the best estimate of American civilian anthropometry.

Dimensions of the head, hand, and foot are apparently much the same in military and civilian populations. Hence, military surveys can serve as sources for data on heads; for hands, Garrett (1971) provided special information, as did White (1982) for feet.

Functional Anthropometry

Classical anthropometric data provide information on static dimensions of the human body in standard postures. However, these data do not describe functional performance capabilities, such as reach capabilities. These are traditionally measured with either the tip of a finger just touching an object, or with the tips of several fingers enclosing a (small) object, or with the whole hand grasping an object. (For other hand-object couplings see Figure 1-11.) The reach measurement usually is thought to originate from the shoulder joint, and encompassing the whole reach envelope around this joint with the arm extended where possible or bent in elbow and wrist when needed.

However, while the shoulder joint as reference point makes some "anatomical sense," this is not a practical origin for the data. Therefore, most reach studies have employed devices that define, by their construction, the reference point for the measures. (Roebuck, Kroemer, and Thomson (1975) give an overview over such devices and their specific features.) Most data on reach envelopes employ the Seat Reference Point (SRP) as origin, which is the point in the mid-sagittal plane where the surfaces of the seatback and of the seatpan meet. Figure 1-12 shows such a method to measure the reach envelope. (Note, however, that the "forward reach" data in Table 1-6 were measured from a wall against which the subject leaned.)

Another topic of applied anthropometry is that of space needs and workplace dimensions for the body in common working postures. Since it is difficult to define working postures which vary from task to task, very few data have been actually measured: Figure 1-13 shows some workplace information for male military personnel. For other populations and for specific work tasks, data can be derived from standard anthropometry (such as in Table 1-6) or must be specifically measured to provide the information needed in a given case, such as described by Roebuck, Kroemer, and Thomson (1975).

VARIABILITY OF ANTHROPOMETRIC DATA

Causes and symptoms of variability in anthropometric data can be divided into three groups: inter-individual variations, intra-individual variations, and secular changes.

Inter-Individual Variations

Inter-individual variations (such as reflected in Table 1-6) are a result of DNA characteristics of which exist about 10^9 possible chromosome combinations. An individual's genetic endorsement determines his/her cellular composition (genotype) and biologically measurable characteristics (phenotype). In addition, the individual's body size is influenced by the environment, e.g., altitude, temperature, sun light, perhaps soil type. Obviously, nutrition also has direct effects with increasing body size (obesity) resulting from over-feeding, and lack of nourishment leading to slenderness. (By the way, how beauty idols change: around 1600 Rubens painted rotund and voluptuous persons as desirable, while in the 1980s the "starvation look" was fashionable.)

1. Finger Touch:
    One finger touches an object without holding it.

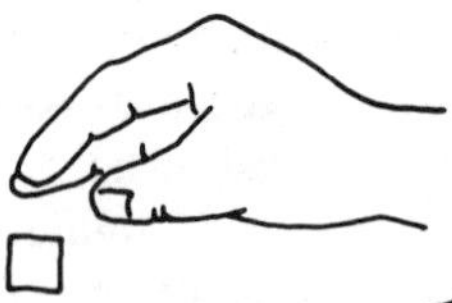

2. Palm Touch:
    Some part of the inner surface of the hand touches the object
    without holding it.

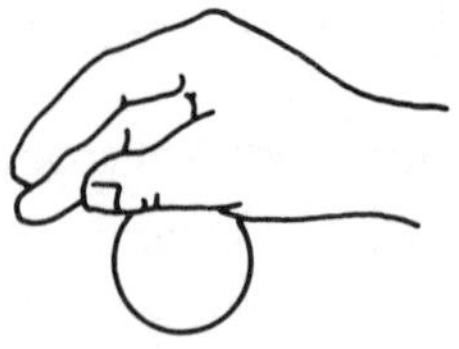

3. Finger Palmar Grip ("Hook Grip"):
    One finger or several fingers hook(s) onto a ridge, or handle.
    This type of finger action is used where thumb counterforce
    is not needed.

4. Thumb-Fingertip Grip ("Tip Grip"):
    The thumb tip opposes one fingertip.

5. Thumb-Finger Palmar Grip ("Pinch Grip"):
    Thumb pad opposes the palmar pad of one finger, or the pads
    of several fingers near the tips. This grip evolves easily from
    coupling #4.

6. Thumb-Forefinger Side Grip (Lateral Grip or
    "Side Pinch"):   Thumb opposes the (radial) side of the
    forefinger.

7. Thumb-Two-Finger Grip ("Writing Grip"):
    Thumb and two fingers (often forefinger and index finger)
    oppose each other at or near the tips.

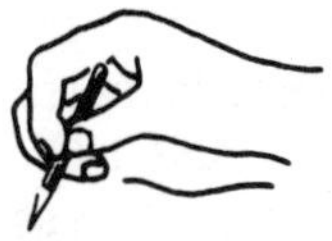

8. Thumb-Fingertips Enclosure ("Disk Grip"):
    Thumb pad and the pads of three or four fingers oppose each
    other near the tips (object grasped does not touch the palm).
    This grip evolves easily from coupling #7.

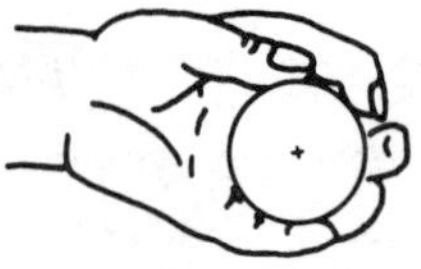

9. Finger-Palm Enclosure ("Enclosure"):
    Most, or all, of the inner surface of the hand is in contact
    with the object while enclosing it.

10. Grasp ("Power Grasp"):
    The total inner hand surface is grasping the (often cylindri-
    cal) handle which runs parallel to the knuckles and generally
    protrudes on one side or both sides from the hand.

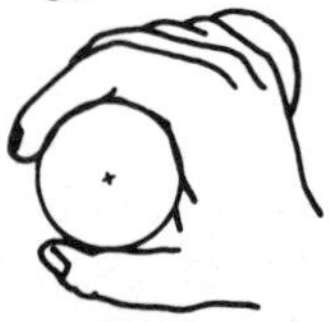

Figure 1-11. Various couplings between hand and object, ranging from a mere
"touch" through "grip" and "enclosure" to the powerful "grasp"

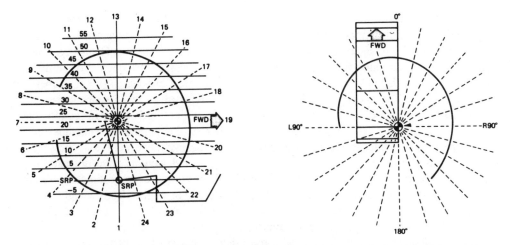

Figure 1-12. Example of the reach envelope measured from the Seat Reference Point SRP in different horizontal planes: shown is the side view (left: height levels in inches) and the top view (right) (from Roebuck, Kroemer, and Thomson 1975).

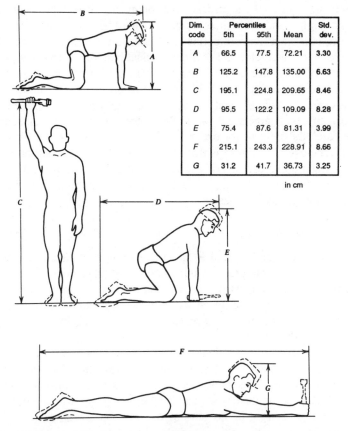

Dim. code	Percentiles		Mean	Std. dev.
	5th	95th		
A	66.5	77.5	72.21	3.30
B	125.2	147.8	135.00	6.63
C	195.1	224.8	209.65	8.46
D	95.5	122.2	109.09	8.28
E	75.4	87.6	81.31	3.99
F	215.1	243.3	228.91	8.66
G	31.2	41.7	36.73	3.25

in cm

Figure 1-13. Space needs of male military personnel for non-standard working postures (from Roebuck, Kroemer, and Thomson 1975).

A variety of "recommended" body weight tables exists, often subdivided for gender, age, and body build. Among the best known tables, used by many physicians, are those prepared in 1942 and since updated by the Metropolitan Life Insurance Company. They all rely on the assumption that desirable weights can be gleaned from population statistics arranged for the dependent variable "mortality" from specific diseases (heart, diabetes, cancer, stroke). The underlying idea was that the weights of 20 to 25 year old persons were "ideal" and should be maintained throughout life, with modifications for gender and body build. Andres (1984) questioned the validity of the population statistics on a number of anthropometric issues: that persons seeking insurance may not represent a random sample of the general population; that heights and weights were not carefully measured; that height and weight measurements by themselves are poor estimates of body obesity; and that no measurements of body build were ever made. Body weight is not consistently correlated with stature: It may be above 0.6 in highly selected groups (e.g., airline stewardesses), about 0.5 in the military, and much lower in the general population.

The population of the U.S. (and of some other countries) is a composite of many different races and ethnic origins. For example, about 100 million U.S. citizens say they have either English or German roots, while Irish ancestry is claimed by approximately 40 million; nearly 21 million claim to be of African descent; about 8 million stem from Mexico, and more than 2 1/2 million are of East Asian origin. Table 1-8 contains more details. There are statistically significant anthropometric differences between groups of various ethnic origins, and while these differences may be of practical importance for the design and use of certain items in defined localities, on a nationwide scale these differences are of no great magnitude and of little practical importance (see, e.g., NASA 1978). The variations in body size within groups are usually much more striking than the average differences between groups. This is also true for differences in body sizes among various professions; "white" and "blue collar" groups are not much different anthropometrically, on a large scale. A similar statement applies to "handedness" (which needs to be carefully defined, for example indicating the preference for writing or for hammering). The overall estimate is that about 10% of all Americans are "left-handed." This may be of some interest to, say, the manufacturers of special hand tools but otherwise does not imply large differences in arm lengths, or arm circumferences, as reflected in nationwide anthropometric tables.

Intra-Individual Variations

Among the intra-individual variables, the effects of aging on anthropometry are rather obvious. During the growing years, stature, weight and other body dimensions increase, which then become relatively stable in early adulthood. With increasing age certain dimensions begin to be reduced (such as body height) while circumferences and the external diameters of bones usually increase. Table 1-9 lists, in approximate numbers, changes in stature with age.

Other examples of intra-individual variations are variations in weight or circumferences associated, e.g., with changes in nutritional and physical activities. (In fact, part of the foregoing discussion on "desirable weights" concerned intra-individual aspects.) Variations in stature during the day are also a case of changes within a person. Immediately after rising from bedrest in the morning one may be several centimeters taller than after a full day "on the feet": this is mainly due to the loss of body fluids from the intervertebral disks as a result of compressive forces generated by gravity and body activities.

Table 1-8. Ancestry claimed by groups of 10^6 U.S. citizens (Source: U.S. Census Bureau Poll 1980).

English	49.6	Japanese	0.8
German	49.2	French Canadian	0.8
Irish	40.2	Slovak	0.8
Afro-American	21.0	Lithuanian	0.7
French	12.9	Ukrainian	0.7
Italian	12.2	Finnish	0.6
Scottish	10.1	Cuban	0.6
Polish	8.2	Canadian	0.5
Mexican	7.7	Korean	0.4
American Indian	6.7	Belgian	0.4
Dutch	6.3	Yugoslavian	0.4
Swedish	4.4	Romanian	0.3
Norwegian	3.5	Asian Indian	0.3
Russian	2.8	Labanese	0.3
Spanish-Hispanic	2.7	Jamaican	0.3
Czech	1.9	Croatian	0.3
Hungarian	1.8	Vietnamese	0.2
Welsh	1.7	Armenian	0.2
Danish	1.5	African	0.2
Puerto Rican	1.4	Hawaiian	0.2
Portuguese	1.0	Dominican	0.2
Swiss	1.0	Columbian	0.2
Greek	1.0	Slovenic	0.1
Austrian	1.0	Iranian	0.1
Chinese	0.9	Syrian	0.1
Filipino	0.8	Serbian	0.1

Table 1-9. Approximate changes in stature with age.

Age (in years)	Change (in cm) Females	Males
1 to 5*	+36	+36
5 to 10	+28	+27
10 to 15	+22	+30
15 to 20	+1	+6
20 to 35**	0	0
35 to 40	-1	0
40 to 50	-1	-1
50 to 60	-1	-1
60 to 70	-1	-1
70 to 80	-1	-1
80 to 90	-1	-1

* Average stature at age 1: females 74 cm, males 75 cm
** Average maximal stature: females 161 cm, males 174 cm

Secular Variations

When looking at medieval armor displayed in a museum one cannot help but notice the apparently small sizes of soldiers centuries ago. In fact, there is some factual and much anecdotal evidence that people are nowadays, on the average, larger than their ancestors. However, "hard" anthropometric information on this development is only available for approximately the last hundred years. From the mid-19th century on, anthropometric surveys were done in reasonably consistent manners on sufficiently large samples. Figure 1-14 provides information on stature for a variety of civilian samples: the over-all trend is apparent. Similarly, Figure 1-15 presents information on military data measured in the U.S. In comparing the two illustrations one is surprised that the military data show virtually no change between the Civil War and World War I but a pronounced increase thereafter. Why stature was seemingly stable for nearly 60 years and then increased rapidly is open to speculation: Perhaps the recruitment of soldiers from the general population was different during the Civil War from the screening process during World War I; or the general population had a massive influx of shorter immigrants; or measurement techniques may have been systematically different. (This brief discussion points out some of the difficulties in comparing time-wise and technically disjointed data.) Nevertheless, the increase observed in stature during the last 50 years is apparently "real." Data from virtually all major surveys in the U.S. and Europe indicate an increase in stature of about 1 cm per decade. (Weight increases were even more dramatic, in the neighborhood of 2 kg for every ten years.)

It is interesting to speculate about the reasons for these increases and to forecast the future development (Roebuck, Smith, and Raggio 1988). One generally accepted explanation is that improvements in living conditions, both hygienic and nutritional, have allowed people to achieve their genetically possible stature better than in earlier times. If this is true one would expect that, nationwide, an "average maximal height" would be approached in the future asymptotically, provided that improvement in living conditions applies to the whole population. Of course, such movements as weight consciousness (if generally practiced over long periods of time) could alter the development of body weight, one anthropometric characteristic.

Altogether, the secular developments of body dimensions are rather small and slow. Hence, for most engineers the changes in body data should have little practical consequences for the design of tools, equipment, and work places, since virtually none are designed to be used over many decades or even centuries. Most products have a relatively short "design life," for which secular changes in anthropometry of the users have no appreciable importance.

The Changing Population

Populations do not remain constant but change in age, health, strength, etc. Their composition changes as well: for example, the work force in the U.S. has today many more women in occupations that used to be dominated by males just a few decades ago. It is estimated that in the mid-nineties two out of three U.S. workers may be female. Occupations have changed drastically.

Computers and service industries are pulling people from traditional workshops and industries. Fewer people are "blue collar" workers in the traditional sense. Life expectancy in the U.S.A. has increased since 1900 by 27 years to 75 years in 1986; in 2020, approximately 18% of all Americans will be 65 years and older. Thus, the convenient assumption of the broad-based "population pyramid" where many young people support a

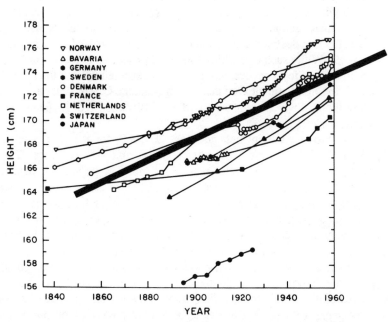

Figure 1-14. Secular increase in stature of young European and Japanese males (adapted from NASA 1978). Heavy line shows the (estimated) apparent trend.

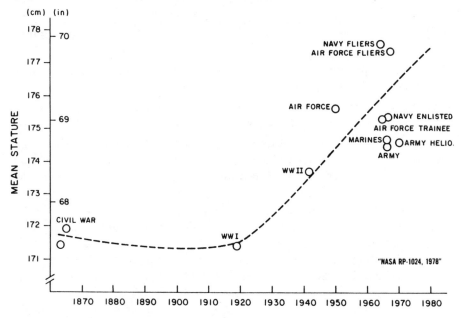

Figure 1-15. Trend of change in stature for U.S. soldiers (NASA 1978).

few old ones is no longer true. In 1971, the median age of the population was almost 28 yrs; in 1980, 30 yrs; in 1986, almost 32 yrs; it is expected to exceed 36 years by the year 2000. Currently, there are more Americans over 65 than there are teenagers.

The total fertility rate declined from 3.7 births per woman in 1960 to 2.5 in 1970 and to about 1.8 in 1985. While such a low birth rate would lead to a reduction in the total U.S. population within just a few decades, immigration is expected to keep the number of U.S. citizens from shrinking. Estimates are that in 100 years the U.S.A. will have a population of about 300 million, of which about 16% will be of Black, 16% of Hispanic, and 10% of Asian ancestry.

Americans are moving within their country. In 1980, for the first time in modern times, the majority of Americans lived in the southern and western states. The sunbelt, prominently California, Florida, and Texas grew more in population than the other regions in the U.S.; migration had reduced the population in the snowbelt. Cities used to be magnets for the rural population; in the 1960s and 1970s the flow was reversed, but in the 1980s some metropolitan areas were again growing. Still, many people seem to move away from the very large cities to smaller, not so crowded communities. However, these trends can change quickly, for example with economic developments.

Such developments require that anthropometric data describing the population in general and, more importantly, depicting certain groups of particular interest to the engineer, be carefully monitored so that their body dimensions can be considered properly in technical designs. For example: a computer workstation that fits a college group in North Dakota well may not be of the correct size and adjustment range for, say, computer operators in southern Texas.

SUMMARY

- Traditional techniques for measuring the human body used a set of simple tools mostly based on grids, rods, and tapes as measuring scales. New developments are using photographic or laser-based three-dimensional methods.

- Traditional measurements are preferably taken to solid identifiable landmarks of the skeleton. Other measurements are — or should be — statistically related to these basic dimensions.

- Most currently available data were measured on (male) soldiers. Data describing civilians are scarce, often derived from military data.

- Body proportions are vastly different among people. Hence, body "types" are not a suitable means to assess body dimensions or capabilities for engineering purposes.

- The construct of the "average person" (or of other single-percentile phantoms) is false and misleading and should be replaced with more realistic statistical approaches, e.g. based on regression equations, and expressed in single-dimension percentiles.

- Changes in body dimensions are due to secular body size development, mobility (within professions, or geographic), age, population, composition, or immigration. Such changes influence the population anthropometry in general only subtly. However, in fitting equipment to a defined population subgroup, the engineer must consider the body dimensions that are specific to the group.

REFERENCES

Andres, R. 1984. Mortality and Obesity: The Rationale for Age-Specific Height-Weight Tables. In *Principles of Geriatric Medicine,* eds. R. Andres, E. L. Bierman, and W. R. Hazzard, pp. 311 –318. New York, NY: McGraw-Hill.

Baughman, L. 1982. Segmentation and Analysis of Sterophotometric Body Surface Data. AFAMRL-TR-81-96. Wright-Patterson AFB, OH: Air Force Aerospace Medical Research Laboratory.

Chaffin, D. B. and Andersson, G. B. J. 1984. *Occupational Biomechanics.* New York, NY: Wiley.

Chapanis, A. (Ed.) 1975. Ethnic Variables in Human Factors Engineering. Baltimore, MD: The John Hopkins University Press.

Churchill, E., Churchill, T., and Kikta, P. 1978. Intercorrelations of Anthropometric Measurements: A Source Book for USA Data. AMRL-TR-77-2. Wright-Patterson AFB, OH: Aerospace Medical Research Laboratory.

Churchill, E., McConville, J. T., Laubach, L. L., and White, R. M. 1971. Anthropometry of U.S. Army Aviators - 1970. TR-72-52-CE. Natick, MA: Clothing and Personal Life Support Equipment Laboratory, United States Army Natick Laboratories.

Department of Defense. 1980. *Anthropometry of U.S. Military Personnel (metric).* DOD-HDBK-743. Washington, D.C.: U.S. Government Printing Office.

Drerup, B. and Hierholzer, E. 1985. Objective Determination of Anatomical Landmarks on the Body Surface: Measurement of the Vertebra Prominens from Surface Curvature. *Biomechanics* 18(6):467–474.

Garrett, J. W. 1971. The Adult Human Hand: Some Anthropometric and Biomechanical Considerations. *Human Factors* 13(2):117–131.

Garrett, J. W. and Kennedy, K. W. 1971. A Collation of Anthropometry. AMRL-TR-68-1. Wright-Patterson AFB, OH: Aerospace Medical Research Laboratories.

Gordon, C. C., Churchill, T., Clauser, C. E., Bradtmiller, B., McConville, J. T., Tebbetts, I., and Walker, R. A. 1989. 1988 Anthropometric Survey of U.S. Army Personnel: Summary Statistics Interim Report. Natick-TR-89/027. Natick, MA: U.S. Army Natick Research, Development and Engineering Center.

Herron, R. E. 1973. Biostereometric Measurement of Body Form. Yearbook of Physical Anthropology 1972, *American Association of Physical Anthropologists* 16:80–121.

Hertzberg, H. T. E. 1968. The Conference on Standardization of Anthropometric Techniques and Terminology. *American Journal of Physical Anthropology* 28(1):1–16.

Kennedy, K. W. 1986. A Collation of United States Air Force Anthropometry. AMRL-TR-85-062. Wright-Patterson AFB, OH: Aerospace Medical Research Laboratory.

Kroemer, K. H. E. 1983. Engineering Anthropometry. In *The Physical Environment at Work,* eds. D. J. Oborne and M. M. Gruneberg, pp. 39–68. London: Wiley.

McConville, J. T., Robinette, K. M., and Churchill, T. 1981. An Anthropometric Data Base for Commercial Design Applications. Final Report, NSF DAR-80 09 861. Yellow Springs, OH: Anthropology Research Project, Inc.

NASA. 1978. *Anthropometric Sourcebook* (3 volumes). NASA Reference Publication 1024. Houston, TX: L. B. J. Space Center, NASA (NTIS, Springfield, VA 22161, Order No. 79 11 734).

Robinette, K. M. and McConville, J. T. 1981. An Alternative to Percentile Models. SAE Technical Paper 810217. Warrendale, PA: Society of Automotive Engineers.

Roebuck, J. A., Kroemer, K. H. E., and Thomson, W. G. 1975. *Engineering Anthropometry Methods.* New York, NY: Wiley.

Roebuck, J., Smith, K., and Raggio, L. 1988. Forecasting Crew Anthropometry for Shuttle and Space Station. In *Proceedings, 32nd Annual Meeting of the Human Factors Society*. pp. 35–39. Santa Monica, CA: Human Factors Society.

Snyder, R. G., Schneider, L. W., Owings, C. L., Reynolds, H. M., Golomb, D. H., and Schork, M. A. 1977. Anthropometry of Infants, Children, and Youths to Age 18 for Product Safety Design. Final Report UM-HSRI-77-17. Ann Arbor, MI. University of Michigan.

Tebbetts, I., Churchill, T., and McConville, J. T. 1980. Anthropometry of Women of the U.S. Army - 1977. TR-80-016. Natick, MA: Clothing, Equipment and Materials Engineering Laboratory, United States Army Natick Research and Development Command.

Wagner, C. 1974. Determination of Finger Flexibility. *European Journal of Applied Physiology* 32:259–278.

Wagner, C. 1988. The Pianist's Hand: Anthropometry and Biomechanics. *Ergonomics*, 31(1):97–131.

White, R. M. 1982. Comparative Anthropometry of the Foot. TR-83/010. Natick, MA: United States Army Natick Research and Development Laboratories.

FURTHER READING

Chaffin, D. B. and Andersson, G. B. J. 1984. *Occupational Biomechanics*. New York, NY: Wiley.

Chapanis, A. ed. 1975. *Ethnic Variables in Human Factors Engineering*. Baltimore, MD: The John Hopkins University Press.

Easterby, R., Kroemer, K. H. E., and Chaffin, D. B. Eds. 1981. *Anthropometry and Biomechanics: Theory and Application*. New York, NY: Plenum.

Gould, S. J. 1981. *The Mismeasure of Man*. New York, NY: Norton.

Hertzberg, H. T. E. 1979. Engineering Anthropology: Past, Present, and Potential. In *The Uses of Anthropology*. No. 11, American Anthropological Association, pp. 184–204.

Lohman, T. G., Roche, A. F., and Martorell, R. Eds. 1988. *Anthropometric Standardization Reference Manual*. Champaign, IL: Human Kinetics.

NASA. 1978. *Anthropometric Sourcebook* (3 volumes). NASA Reference Publication 1024, Houston, TX: NASA (NTIS, Springfield, VA 22161, Order No. 79 11 734).

Pheasant, S. 1986. *Bodyspace: Anthropometry, Ergonomics and Design*. Philadelphia, PA: Taylor and Francis.

Roebuck, J. A., Kroemer, K. H. E., and Thomson, W. G. 1975. *Engineering Anthropometry Methods*. New York, NY: Wiley.

APPENDIX A: CORRELATION TABLE

The following Table 1-10 was compiled from data on male and female U.S. soldiers contained in the publications by Churchill, Churchill, and Kikta (1978); Churchill, McConville, Laubach, and White (1971); and Tebbetts, Churchill, and McConville (1980). It shows the coefficients of correlation between 19 body dimensions important for design purposes

Standing heights:	acromion, cervicale, crotch
Sitting heights:	sitting, eye, knee, popliteal
Sitting depths, breadths:	buttock-knee, popliteal, hip
Forward Reach:	thumb-tip
Circumferences:	chest, waist
Hand Dimensions:	breadth, circumference, length
Head Dimensions:	breadth, circumference, length

and 9 body dimensions selected for their predictive powers, i.e., having high correlations with certain important design variables. Listed for each variable are the correlations for three male samples (top 3 rows), followed by the data for two female samples.

Table 1-10 indicates generally high correlations within heights, within certain circumferences, within measurements on the hand, and within some head dimensions.

Stature is a valuable predictor for many selected design variables, i.e., (standing) acromion, cervicale and crotch height, and for (sitting) height, and eye, knee, popliteal height; even buttock-popliteal length, thumb-tip reach and hand length are rather well correlated. Waist height is a good predictor for trunk heights and for leg lengths, while weight is primarily related to trunk measurements contained in the table. These statements hold true, with only minor shifts, for male and female soldiers.

More detailed correlations within highly specific military samples were compiled by NASA (1978) and Roebuck, Kroemer, and Thomson (1975).

Table 1-10. Correlations between (19) design dimensions and (9) predictor dimensions.

Design Dimensions	Stature	Bitragion Breadth	Head Length	Heel-Ankle Circumference	Hip Breadth, standing	Palm Length	Shoulder Circumference	Waist Height	Weight
Acromion Height, standing	.960**	.218	.260	.599	.397	.575	.306	.911**	.479
	.958**	.207	.224	.622*	.448	.525	.344	.906**	.554
	.959**	.184	.228	.641*	.425	.527	.304	.909**	.480
	.959**	.237	.298	na	.350	na	.350	.919**	.552
	.968**	na	.336	.618*	na	.574	.354	.891**	.567
Cervicale Height, standing	.977**	.210	.263	.611*	.402	.586	.324	.929**	.482
	.977**	.185	.217	.630*	.427	.526	.331	.931**	.529
	.978**	.172	.237	.655*	.420	.545	.308	.919**	.476
	.977**	.221	.299	na	.349	na	.334	.927**	.540
	.979**	na	.320	.599	na	na	.375	.894**	.563
Crotch Height, standing	.839**	.075	.191	.482	.196	.538	.141	.887**	.252
	.856**	.082	.170	.517	.283	.484	.212	.905**	.359
	.857**	.082	.196	.601*	.229	.549	.175	.929**	.312
	.849**	.174	.280	na	.225	na	.264	.909**	.430
	.861**	na	.320	.599	na	.629*	.247	.882**	.402
Sitting Height	.778**	.195	.270	.487	.403	.388	.296	.537	.434
	.786**	.158	.244	.495	.376	.367	.291	.580	.467
	.732**	.196	.227	.404	.422	.275	.289	.476	.412
	.801**	.204	.275	na	.371	na	.312	.607*	.091
	.767**	na	.261	.350	na	.304	.253	.570	.421
Eye Height, sitting	.753**	.173	.233	.461	.388	.374	.281	.518	.411
	.738**	.121	.193	.456	.346	.341	.260	.544	.412
	.709**	.170	.184	.383	.400	.273	.265	.463	.380
	.737**	.170	.229	na	.371	na	.287	.562	.448
	.738**	na	.220	.331	na	.279	.235	.551	.399
Knee Height, sitting	.878**	.216	.258	.618*	.344	.612*	.331	.903**	.460
	.882**	.193	.221	.665*	.428	.539	.358	.904**	.539
	.873**	.177	.253	.738**	.420	.583	.373	.892**	.524
	na	na	na	na	na	na	na	na	na
	.857**	na	.362	.694*	na	.647*	.363	.850**	.546
Popliteal Height, sitting	.808**	.036	.170	.427	.030	.540	.020	.832**	.090
	.841**	.106	.175	.529	.221	.513	.182	.883**	.299
	.830**	.079	.206	.615*	.203	.569	.176	.888**	.289
	.728**	.149	.253	na	.181	na	.230	.762**	.370
	.847**	na	.326	.622*	na	.651*	.252	.853**	.401
Buttock-Knee Length, sitting	.801**	.201	.244	.567	.480	.504	.434	.844**	.582
	.760**	.229	.208	.586	.550	.454	.435	.790**	.636*
	.766**	.224	.249	.682*	.554	.487	.493	.795**	.662*
	.769**	.245	.296	na	.527	na	.482	.809**	.694*
	.761**	na	.342	.671*	na	.539	.506	.748**	.720**
Buttock-Popliteal Length, sitting	.684*	.145	.303	.131	.341	.415	.284	.737**	.397
	.686*	.161	.157	.499	.506	.374	.394	.718**	.565
	.706**	.160	.226	.615*	.491	.446	.430	.748**	.586
	.653*	.197	.249	na	.426	na	.391	.710**	.565
	na	na	na	na	na	na	na	na	na
Thumb-Tip Reach	.627*	.179	.223	.404	.204	.523	.271	.608*	.336
	.676*	.190	.179	.503	.327	.486	.304	.675*	.414
	.676*	.159	.187	.562	.325	.509	.301	.681*	.395
	.646*	.210	.220	na	.252	na	.312	.652*	.433
	na	na	na	na	na	na	na	na	na

Listed for each dimension:
1st line: U.S. Army Flyers, male
2nd line: U.S. Air Force Officers, male
3rd line: U.S. Air Force Trainees, male
4th line: U.S. Air Force Woman
5th line: U.S. Army Women

na: (data) not available
*:   above 0.6
**:   above 0.7

Table 1-10 (continued)

Design Dimensions	Stature	Bitragion Breadth	Head Length	Heel-Ankle Circumference	Hip Breadth, standing	Palm Length	Shoulder Circumference	Waist Height	Weight
					Predictor Dimensions				
Hip Breadth, Sitting	.331	.344	.208	.522	.891**	.182	.723**	.289	.832**
	.372	.312	.159	.512	.903**	.231	.669*	.295	.855**
	.360	.364	.202	.496	.921**	.162	.713**	.210	.875**
	.348	.266	.183	na	na	na	.606*	.318	.770**
	na	na	na	na	na	na	na	na	na
Chest Circumference	.198	.408	.229	.466	.706**	.162	.883**	.193	.878**
	.240	.398	.196	.466	.647*	.217	.822**	.203	.832**
	.246	.363	.214	.476	.721**	.172	.872**	.143	.861**
	.257	.309	.204	na	.551	na	.810**	.238	.799**
	.239	na	.176	na	na	na	na	na	na
Waist Circumference	.180	.377	.186	.430	.769**	.115	.778**	.171	.890**
	.224	.349	.158	.424	.724**	.154	.720**	.142	.856**
	.257	.377	.189	.447	.816**	.122	.764**	.117	.886**
	.279	.353	.149	na	.600	na	.775**	.238	.824**
	.208	na	.145	na	.571	na	na	.061	.787**
Hand Breadth	.441	.303	.267	.565	.300	.439	.351	.383	.423
	.409	.239	.209	.568	.329	.420	.378	.360	.448
	.390	.249	.259	.561	.348	.465	.452	.329	.485
	.380	.275	.241	na	.223	na	.355	.340	.417
	.433	na	.326	na	na	na	na	na	na
Head Circumference	.392	.258	.274	.612*	.378	.418	.440	.356	.510
	.412	.318	.231	.616*	.371	.422	.427	.365	.510
	.442	.279	.276	.626*	.387	.394	.502	.380	.539
	.365	.310	.236	na	.264	na	.424	.321	.495
	.448	na	.314	na	na	na	na	na	na
Hand Length	.661*	.189	.261	.572	.254	.859**	.250	.648*	.339
	.651*	.177	.246	.618*	.309	.844**	.258	.642*	.389
	.624*	.165	.243	.626*	.252	.857**	.288	.632*	.353
	.601*	.229	.316	na	.196	na	.289	.603*	.383
	na	na	na	na	na	na	na	na	na
Head Breadth	.065	.610*	.066	.176	.200	.052	.255	.035	.272
	.132	.622*	.058	.215	.227	.121	.244	.088	.305
	.088	.633*	.103	.190	.289	.036	.328	.016	.348
	.136	.586	.115	na	.188	na	.252	.122	.290
	.154	na	.161	.241	na	na	.302	.115	.317
Head Circumference	.265	.320	.694*	.330	.234	.175	.318	.202	.355
	.294	.424	.778**	.365	.288	.233	.327	.232	.412
	.258	.357	.809**	.371	.320	.190	.392	.182	.446
	.331	.408	.692*	na	.265	na	.330	.306	.403
	.364	na	.796**	.447	na	na	.352	.319	.409
Head Length	.286	.145	--	.327	.180	.206	.240	.221	.293
	.249	.126	--	.286	.166	.196	.201	.208	.261
	.262	.095	--	.334	.188	.205	.260	.216	.308
	.318	.158	--	na	.183	na	.248	.297	.304
	.370	na	--	.388	na	na	.246	.314	.317

Listed for each dimension:
- 1st line: U.S. Army Flyers, male
- 2nd line: U.S. Air Force Officers, male
- 3rd line: U.S. Air Force Trainees, male
- 4th line: U.S. Air Force Woman
- 5th line: U.S. Army Women

na: (data) not available
*:   above 0.6
**:   above 0.7

APPENDIX B: GLOSSARY OF ANATOMICAL AND ANTHROPOMETRIC TERMS

abduct — to move away from the body or one of its parts; opposed to adduct.

acromion — the most lateral point of the lateral edge of the scapula. Acromial height is usually equated with shoulder height.

adduct — to move towards the body; opposed to abduct.

anterior — pertaining to the front of the body; opposed to posterior.

axilla — the armpit.

biceps brachii — the large muscle on the anterior surface of the upper arm, connecting the scapula with the radius.

biceps femoris — a large posterior muscle of the thigh.

brachialis — muscle connecting the mid-humerus with the ulna.

buttock protrusion — the maximal posterior protrusion of the right buttock.

carpus — the wristbones, collectively.

cervicale — the protrusion of the spinal column at the base of the neck caused by the tip of the spine of the 7th cervical vertebra.

clavicle — the "collarbone" linking the scapula with the sternum.

condyle — articular prominence of a bone.

coronal plane — any vertical plane at right angles to the midsagittal plane (same as frontal plane).

dactylion — the tip of the middle finger.

distal — the end of a body segment farthest from the head, opposed to proximal.

dorsal — pertaining to the back, also to the top of hand or foot, opposed to palmar, plantar, and ventral.

epicondyle — the bony eminence at the distal end of the humerus, radius, and femur.

extend — to move adjacent segments so that the angle between them is increased, as when the leg is straightened; opposed to flex.

external — away from the central long axis of the body; the outer portion of a body segment.

femur — the thigh bone.

flex — to move a joint in such a direction as to bring together the two parts which it connects, as when the elbow is bent; opposed to extend.

Frankfurt Plane — the standard horizontal plane for orientation of the head. The plane is established by a line passing through the right tragion (approximate ear hole) and the lowest point of the right orbit (eye socket), with both eyes on the same level.

frontal plane — same as coronal plane.

glabella — the most anterior point of the forehead between the brow ridges in the midsagittal plane.

glenoid cavity — depression in the scapula below the acromion into which fits the head of the humerus, forming the "shoulder joint".

gluteal furrow — the furrow at the juncture of the buttock and the thigh.

humerus — the bone of the upper arm.

iliac crest — the superior rim of the pelvic bone.

illium — see pelvis.

inferior — below, lower, in relation to another structure.

inseam — a term used in tailoring to indicate the inside length of a sleeve or trouser leg. It is measured on the medial side of the arm or leg.

internal — near the central long axis of the body; the inner portion of a body segment.

ischium — the dorsal and posterior of the three principal bones that compose either half of the pelvis.

knuckle — the joint formed by the meeting of a finger bone (phalanx) with a palm bone (metacarpal).

lateral — lying near or toward the sides of the body; opposed to medial.

malleolus — a rounded bony projection in the ankle region. The tibia has such a protrusion on the medial side, and the fibula one on the lateral side.

medial — lying near or toward the midline of the body; opposed to lateral.

metacarpal — pertaining to the long bones of the hand between the carpus and the phalanges.

mid-sagittal plane — the vertical plane which divides the body (in the anatomical position) into right and left halves.

olecranon — the proximal end of the ulna.

omphalion — the center point of the navel.

orbit — the eye socket.

palmar — pertaining to the palm (inside) of the hand; opposed to dorsal.

patella — the kneecap.

pelvis — the bones of the "pelvic girdle" consisting of illium, pubic arch and ischium which compose either half of the pelvis.

phalanges — the bones of the fingers and toes (singular, phalanx).

plantar — pertaining to the sole of the foot.

popliteal — pertaining to the ligament behind the knee or to the part of the leg behind the knee.

posterior — pertaining to the back of the body; opposed to anterior.

proximal — the end of a body segment nearest the head; opposed to distal.

radius — the bone of the forearm on its thumb-side.

sagittal — pertaining to the midsagittal plane of the body, or to a parallel plane.

scapula — the shoulder blade.

scye — a tailoring term to designate the armhole of a garment. Refers here to landmarks which approximate the lower level of the axilla.

sphyrion — the most distal extension of the tibia on the medial side under the malleolus.

spine — the stack of vertebrae.

spine (or spinal process) of a vertebra — the posterior prominence.

sternum — the breastbone.

stylion — the most distal point on the styloid process of the radius.

styloid process — a long, spinelike projection of a bone.

sub — a prefix designating below or under.

superior — above, in relation to another structure; higher.

supra — prefix designating above or on.

tarsus — the collection of bones in the ankle joint.

tibia — the medial bone of the leg (shin bone).

tibiale — the uppermost point of the medial margin of the tibia.

tragion — the point located at the notch just above the tragus of the ear.

tragus — conical eminence of the auricle (pinna, external ear) in front of the ear hole.

transverse plane — horizontal plane through the body, orthogonal to the sagittal and the coronal planes.

triceps — the muscle of the posterior upper arm.

trochanterion — the tip of the bony lateral protrusion of the proximal end of the femur.

tuberosity — a (large) rounded prominence on a bone.

ulna — the bone of the forearm on its little-finger side.

umbilicus — depression in abdominal wall where the umbilical cord was attached to the embryo.

ventral — pertaining to the anterior side of the trunk.

vertebra — a bone of the spine.

vertex — the top of the head.

CHAPTER 2

THE SKELETAL SYSTEM

OVERVIEW

The skeletal system of the human body is composed of some 200 bones, of their articulations, and of connective tissue. They all consist of special cells imbedded in an extracellular matrix of fibers in a ground substance. Bones provide the stable framework for the body. Ligaments connect bones in their articulations, while tendons connect muscle with bone. The spinal column is of great concern to the engineer because it is the locus of many overexertion injuries at work.

The Model

Links between the joints are provided by the long bones. They provide the basic framework for the body and the lever arms to which tendons attach for muscle torque exertion about the joints. Mobility at the joints is limited by their design, by cartilage, and by muscles.

TISSUE COMPONENTS

Bones and connective tissue are composed of cells embedded in a fiber matrix and ground substance. The cells in cartilage are called chondrocytes, while fibroblasts are found in loose tissue, such as skin, tendons, ligaments, and adipose tissue. The cells are enclosed by the so-called extracellular matrix, which contains two different kinds of fibers: collagen fibers (subdivided into three types) have high tensile strength and resist deformation, particularly stretch, while elastic fibers elongate. The ground substance has large molecules (proteoglycan) with a protein core, and proteins (glycoproteins), calcium (in bone), lipids, and water. The actual composition of tissues differs in bone, articulations, connective tissue, and muscles.

Bones

The main function of human skeletal bone — see Figure 2-1 — is to provide the internal framework for the whole body; without its support, the entire body would collapse into a heap of soft tissue. One distinguishes between flat axial (appendicular) bones, such as in the skull, sternum and ribs, and the pelvis; and long more or less cylindrical bones. The long bones consist of a shaft (diaphysis) which, at each end, broadens into an epiphysis which forms part of the articulation to the adjacent bone. Both flat and long bones consist of compact (cortical) and spongy (cancellous) material, with the latter mostly found in the epiphyses where it transfers the load experienced from the next bone. Furthermore, bones provide shells to protect body portions; for example, the rib cage protects lungs and heart. Long hollow bones also provide the room for bone marrow which serves as a blood factory. Finally, bones provide a reservoir of calcium and phosphorus. Mechanically, bones are the lever arms at which muscles pull about the articulations — see Chapter 5.

While bone is firm and hard, and thus can resist high strain, it still has certain elastic properties, particularly in childhood when mineralization (ratio of the contents of inorganic material, mostly calcium, to organic substance) is relatively low (about 1:1). In contrast, bones of the elderly are highly mineralized, and therefore more brittle (7:1). Bone develops from a soft material in the earliest childhood into compact fibrous material with a hard outer shell and a spongy inner section. Growth takes place until about 30 years of age, whereafter bones usually become more and more osteoporotic which means a decrease in mass, a reduction in the thickness of the outer layer accompanied by an increase in the outer diameter of long bones, bringing about "hollowing" of their core.

Bone cells are nourished through canals carrying blood vessels and tubules. Bone is continuously resorbed and rebuilt throughout one's life; local strain encourages growth, disuse resorption ("Wolff's Law"). Adult bone material consists of osteocytes — cells with long branching processes that occupy cavities (lacumas) in a matrix of densely packed collagenous fibers which in turn are in an amorphous ground substance (called cement) with a high calcium phosphate content.

Connective Tissues

Dense connective tissues, mostly composed of collagen fibers, are called *ligaments* when they connect bones, *tendons* when they connect muscle with bone, and *fascia* when they wrap organs or muscles. The wrapping fascial tissue of muscle fibers condenses at the ends of the muscle to tendons which are usually encapsuled by a fibrous tissue, sheaths, which allows a gliding motion of the tendon against surrounding materials through an inner lining, synovium, which produces a viscous fluid, synovia, that reduces

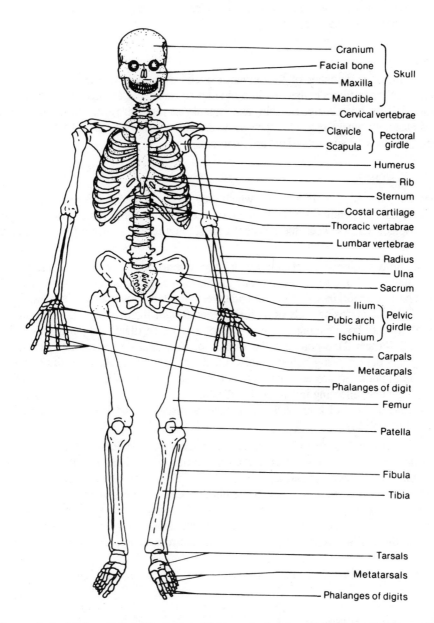

Figure 2-1. Major skeletal bones (with permission from Weller and Wiley 1979).

friction. Particularly in the wrist and fingers, rings or bands of ligaments keep the tendon sheaths close to underlying bones, acting as guides or pulleys for the pulling actions of the tendons and their muscles.

Cartilage

Another major component of the human skeletal system is cartilage, a translucent material of collagen fibers embedded in a binding substance. Cartilage supports moderate strain, is firm but elastic, flexible, and capable of rapid growth. In the human, it supplies elasticity where required, such as at the end of the ribs, as disks between the vertebrae, and as joint surfaces of the articulations.

ARTICULATIONS

Bones are connected by articulations which may be considered the bearing surfaces of adjacent bones. Depending on their structure, these joints allow none, little, or much relative displacement between the adjacent bones. Figure 2-2 shows different types of joints.

Fibrous joints are junctions of bones which are fastened together by fibrous tissue; no tissue separates the bone surfaces. Fibrous joints, such as in the skull, allow no appreciable motion. *Cartilaginous* joints provide very limited movement: here, articular cartilage lines the opposing ends of the bones, and a tight ligament covers the joint and extends along both bone endings. The spinal column has cartilaginous joints, in which a flat disk of fibrocartilage connects the linings of two opposing vertebral bodies. In *synovial* joints the opposing bones are also lined by articular cartilages at the opposing bone surfaces, but these are separated by a space, and the joint is loosely encapsuled by an elastic ligament, allowing much relative displacement of the bones. In some synovial joints, a fibrocartilage disk or wedge (e.g., the meniscus in the knee) is in the space allowing high mobility.

Articular cartilage has no blood vessels. Its nourishment is achieved through direct communication between the cavities of the end portion of the bone (epiphyses) and the basal portions of the articular cartilage. Also, synovial fluid may be exchanged between cartilage and articular space. Synovial membranes can secrete synovia, which acts as lubricant. A strained joint, such as the knee joint in running, can show an increase in cartilage thickness of ten or more percent, brought about in a few minutes by fluid seeping into the cartilage from the underlying bone marrow cavity. This helps to "fill and

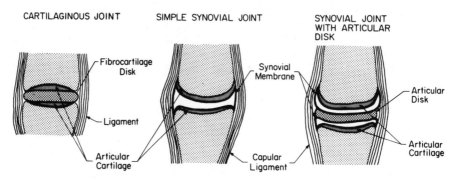

Figure 2-2. Types of body joints (modified from Astrand and Rodahl 1977 and from Chaffin and Anderson 1984).

smooth out" the space between the opposing bones of the joint, reducing the danger of local pressures and damage. Similarly, fluid "seeps" in the spinal disks when they are not compressed, e.g., during a night's bed rest, and makes them more pliable, which may explain disk deformations experienced as "early morning back pains."

Mobility

Most articulations of the human body (excepting the spine) belong to the synovial group. Movements in such joints can be a simple gliding movement, or rather complex movements in several planes. Examples are the rotatory pronation and supination of the hand achieved by relative gliding-twisting of radius and ulna within the forearm (one degree of freedom); the hinge-joint type angular movement of the forearm around the upper arm in the elbow joint (also one degree of freedom); or the complex spacial circumduction which can be performed (with three degrees-of-freedom) in the shoulder or hip. The motions themselves are limited by the form of the bony surfaces and by the tension generated by ligaments and muscles.

Mobility, incorrectly also called flexibility, indicates the range of motion that can be achieved in body articulations. It is usually measured as the difference between the smallest and largest angles enclosed by neighboring body segments, about their common point of rotation, i.e., the articulation. Figure 2-3 illustrates possible motions in major joints. Unfortunately, these data are not as clear and clean as they appear: quite often, the actual point of rotation moves with the motion; for example, the geometric location of the axis of the knee joint moves slightly while the thigh and the lower leg rotate about it. Furthermore, the actual arms of the rotation angle are not well defined: they may be established by straight lines between the center of rotation of the enclosed articulation to the point of rotation of the next articulation (in the example of the knee joint, running distally to the ankle, and proximately to the center of the hip joint); or they may simply be estimated as representing the *mid*-axes of the adjacent body segments.

The range of motion depends on gender, training, and age; and on whether only active contraction of the relevant muscles is used, or if gravity or perhaps even a second person helps to achieve extreme locations. Finally, of course, the mobility is different in different planes if the articulation in question has more than one degree of freedom.

A study by Staff (1983) has provided reliable information about voluntary (unforced) mobility in major body joints. This study was done on 100 females (ages 18 through 35 years) and carefully controlled to resemble an earlier study by Houy (1982) on 100 male subjects. On each person, only one measurement was taken for each motion, recorded by an electronic bubble goniometer. All subjects were asked to move their "limbs only as far as comfortably possible" using (if applicable) the dominant limb; 85% claimed to be right-dominant. The results are compiled in Table 2-1. Of the 32 measurements, 24 showed significantly more mobility by females than by males; men had larger mobility only in ankle flexion and wrist abduction. This finding confirms earlier studies that also showed larger motion capability, in most cases, exhibited by women. Given the careful and consistent conduct of the two studies, one may assume that the data in Table 2-1 are a true reflection of the adult U.S. population within the working age span; mobility is only slightly reduced (less than about $10°$ in the extreme positions) for most body joints as one reaches the 6th decade.

The control of motions in the joints is mostly effected by the muscles that span them. Moveable joints have nervous connections with the muscles that act on them, establishing local reflex arcs which prevent overextensions. Four types of nerve endings exist in the joints. Three of them terminate in specialized organs, called Ruffini organs. They

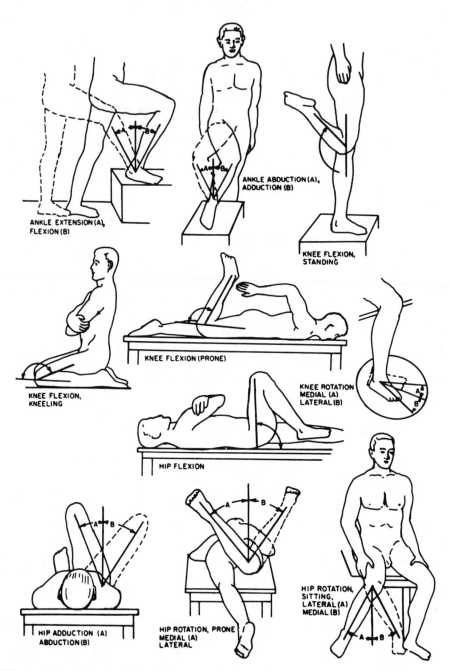

Figure 2-3a. Maximal displacements in body joints (from Van Cott and Kinkade 1972).

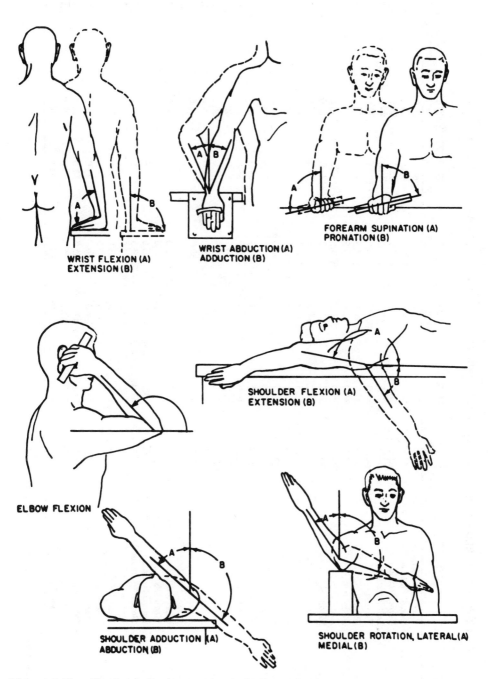

Figure 2-3b. Maximal displacements in body joints (from Van Cott and Kinkade 1972).

Table 2-1. Comparison of mobility data for females and males in degrees (adapted from Staff 1983).

JOINT	MOVEMENT	5TH PERCENTILE		50 PERCENTILE		95TH PERCENTILE		DIFFERENCE*
		FEMALE	MALE	FEMALE	MALE	FEMALE	MALE	FEMALE > MALE
Neck	Ventral Flexion	34.0	25.0	51.5	43.0	69.0	60.0	+8.5
	Dorsal Flexion	47.5	38.0	70.5	56.5	93.5	74.0	+14.0
	Right Rotation	67.0	56.0	81.0	74.0	95.0	85.0	+7.0
	Left Rotation	64.0	67.5	77.0	77.0	90.0	85.0	NS
Shoulder	Flexion	169.5	161.0	184.5	178.0	199.5	193.5	+6.5
	Extension	47.0	41.5	66.0	57.5	85.0	76.0	+8.5
	Adduction	37.5	36.0	52.5	50.5	67.5	63.0	NS
	Abduction	106.0	106.0	122.5	123.5	139.0	140.0	NS
	Medial Rotation	94.0	68.5	110.5	95.0	127.0	114.0	+15.5
	Lateral Rotation	19.5	16.0	37.0	31.5	54.5	46.0	+5.5
Elbow-Forearm	Flexion	135.5	122.5	148.0	138.0	160.5	150.0	+10.0
	Supination	87.0	86.0	108.5	107.5	130.0	135.0	NS
	Pronation	63.0	42.5	81.0	65.0	99.0	86.5	+16.0
Wrist	Extension	56.5	47.0	72.0	62.0	87.5	76.0	+10.0
	Flexion	53.5	50.5	71.5	67.5	89.5	85.0	+4.0
	Adduction	16.5	14.0	26.5	22.0	36.5	30.0	+4.5
	Abduction	19.0	22.0	28.0	30.5	37.0	40.0	-2.5
Hip	Flexion	103.0	95.0	125.0	109.5	147.0	130.0	+15.5
	Adduction	27.0	15.5	38.5	26.0	50.0	39.0	+12.5
	Abduction	47.0	38.0	66.0	59.0	85.0	81.0	+7.0
	Medial Rotation (Prone)	30.5	30.0	44.5	46.0	58.5	62.5	NS
	Lateral Rotation (Prone)	29.0	21.5	45.5	33.0	62.0	46.0	+12.5
	Medial Rotation (Sitting)	20.5	18.0	32.0	28.0	43.5	43.0	+4.0
	Lateral Rotation (Sitting)	20.5	18.0	33.0	26.5	45.5	37.0	+6.5
Knee	Flexion (Standing)	99.5	87.0	113.5	103.5	127.5	122.0	+10.0
	Flexion (Prone)	116.0	99.5	130.0	117.0	144.0	130.0	+13.0
	Medial Rotation	18.5	14.5	31.5	23.0	44.5	35.0	+8.5
	Lateral Rotation	28.5	21.0	43.5	33.5	58.5	48.0	+10.0
Ankle	Flexion	13.0	18.0	23.0	29.0	33.0	34.0	-6.0
	Extension	30.5	21.0	41.0	35.5	51.5	51.5	+5.5
	Adduction	13.0	15.0	23.5	25.0	34.0	38.0	NS
	Abduction	11.5	11.0	24.0	19.0	36.5	30.0	+5.0

*Listed are only differences at the 50th percentile, and if significant ($\alpha < 0.5$)

provide information about changes in joint position, speed of movement, and the actual positioning of the joint. The two first Ruffini organs are located in the joint capsule, the other one in the ligament. The fourth receptor is a free branching nerve ending in pain-sensitive fibers. Synovial membranes and joint cartilage do not have nerve receptors.

THE SPINAL COLUMN

The spinal column is a particularly interesting and complex structure. As shown in Figure 2-4, it consists of 25 bony components: seven cervical, twelve thoracic, and five lumbar vertebrae, and the sacrum with the coccyx which consist of fused groups of rudimentary bones. These components are held together by cartilaginous joints in which the main bodies of the vertebrae are connected by fibrocartilage disks. However, each vertebra also has two protuberances extending backwards-upwards, the superior articulation processes (see Figures 2-5 and 2-6) which end in rounded surfaces fitting into cavities on the underside of the next-higher vertebra. (The synovial facet joints are covered with sensitive tissue.) Hence, the vertebrae rest upon each other in three joints of two different kinds. This complex rod, sustaining the trunk, is kept in delicate balance by ligaments and particularly by muscles running along the posterior side of the spinal column and located along the sides and front of the trunk.

The nucleus pulposus of the spinal disk has no blood supply and no sensory nerves. It is nourished through the disk as a result of osmotic pressure, gravitational force, and the pumping effects of spine movement.

The spinal column is very often the site of overexertion. This may manifest itself by muscular and cartilaginous strains and sprains, and/or in deformation of or even damage to the disks or to the vertebrae. With such injuries currently being a very frequent and very costly problem in U.S. industries, much research has been directed at the causes and mechanisms involved (see, e.g., White and Panjabi 1978.) Since no detailed discussion of problems is possible here (or in Chapter 5), it must suffice to state that many problems seem to be associated with the muscles or cartilage stabilizing the bony stack of vertebrae. Uncoordinated pull of the muscles on this column, particularly when associated with high and/or asymmetric external loads may (directly or indirectly, acutely or cumulatively, often in unknown and not reconstructable manners) lead to various backstrains and back injuries (Pope, Frymoyer, and Andersson 1984). Some but not all occasions for such overexertion injuries can be avoided by ergonomic design of work place, work equipment, and work task (Kroemer 1988). Even for persons with persistent back pains, work places and procedures can be engineered to allow performance of suitable physical work (Rodgers 1985).

SUMMARY

The skeletal system with its bones, joints, and connective tissue is, from the engineering point of view, highly complex. It allows a varying but well controlled range of motion, even under high external loads. Its internal maintenance is highly variable, reduced with age but increased by loading, which strengthens strained bones and lubricates loaded joints. Its control is accomplished by nervous reflex feedback and by excitation signals from the central nervous system to muscles which stabilize and load bones and joints. The spinal column, in particular, has been and is the object of many biomechanical studies.

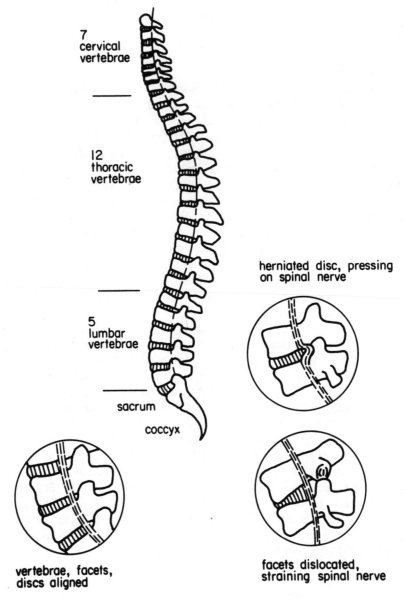

7
cervical
vertebrae

_____

12
thoracic
vertebrae

herniated disc, pressing
on spinal nerve

5
lumbar
vertebrae

_____

sacrum

coccyx

vertebrae, facets,
discs aligned

facets dislocated,
straining spinal nerve

Figure 2-4. Scheme of the human spinal column.

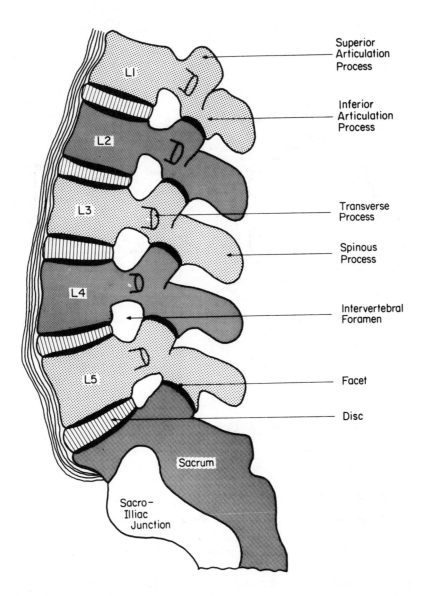

Figure 2-5. Scheme of the lumbar section of the spinal column (bearing surfaces in heavy lines).

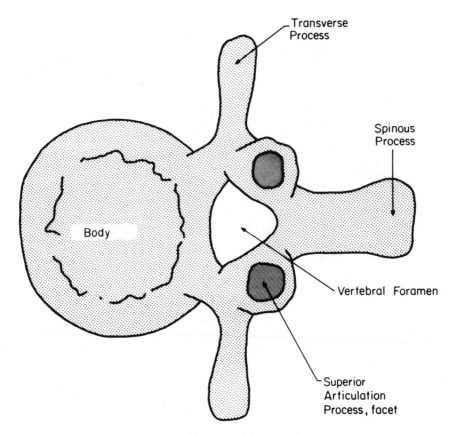

Figure 2-6. Scheme of a vertebra.

REFERENCES

Chaffin, D. B. and Andersson G. B. J. 1984. *Occupational Biomechanics*. New York, NY: Wiley.

Kroemer, K. H. E. 1988. *Ergonomics Manual for Manual Material Handling*. 5th revised edition. Radford, VA: Author.

Pope, M. H., Frymoyer, J. W., and Andersson, G. (eds.). 1984. *Occupational Low Back Pain*. Philadelphia, PA: Praeger.

Staff, K. R. 1983. A Comparison of Range of Joint Mobility in College Females and Males. Industrial Engineering. College Station, TX: Texas A&M University.

Rodgers, S. H. 1985. *Working with Backache*. Fairport, NY: Perinton.

Van Cott, H. P. and Kinkade, R. G. 1972. Human Engineering Guide to Equipment Design. Washington, DC: U.S. Government Printing Office.

Weller, H. and Wiley, R. L. 1979. *Basic Human Physiology*. New York, NY: Van Nostrand Reinhold.

White, A. A. and Panjabi, M. M. 1978. *The Clinical Biomechanics of the Spine*. Philadelphia, PA: Lippincott.

FURTHER READING

Chaffin, D. B. and Andersson G. B. J. 1984. *Occupational Biomechanics.*   New York, NY: Wiley.

Currey, J. 1984. *The Mechanical Adaptations of Bone.* Princeton, NJ: Princeton University Press.

Rodgers, S. H. 1985. *Working with Backache.* Fairport, NY: Perinton.

White, A. A. 1983. *Your Aching Back.* Toronto: Bantam.

Winter, D. A. 1979. *Biomechanics of Human Movement.* New York, NY: Wiley.

CHAPTER 3

SKELETAL MUSCLE

OVERVIEW

Skeletal muscles move body segments with respect to each other. Shortening is the only active function of the muscle. This contraction is controlled by the central nervous system at the motor endplates, where "rate" or "recruitment" coded signals stimulate muscle components to shorten either statically (in isometric or isotonic efforts) or dynamically (isokinetically or isoinertially). Various methods and techniques are available for assessing muscular control and strength.

The Model

Skeletal muscles connect body links across their joint. Muscles are usually arranged in "functional pairs" so that the contracting muscle is counteracted by its opponent. The actual contraction force within a living human muscle depends on internal and external variables.

ARCHITECTURE

The human body has three types of muscle: striated (skeletal), cardiac, and smooth muscle. Striated muscle (also called voluntary muscle) is responsible for locomotion and posture; contractions of the heart's cardiac muscle move blood through the vascular system; and smooth muscle helps to maintain the internal control of the body through contraction and dilation of vascular (in the walls of the blood vessels) and visceral (internal organ) muscle tissues. The electrical stimulation and contraction mechanisms of cardiac muscle and smooth muscle are similar, while the anatomic and physiologic characteristics of cardiac muscle closely resemble those of striated skeletal muscle.

For the engineer, skeletal muscles are of primary interest since they move the segments of the human body under voluntary control and generate energy for exertion to outside objects. There are several hundred skeletal muscles in the human body, known by their Latin names. Many actually consist of bundles of muscles (*fasciculi*) each of which is wrapped, as is the total muscle, in connective tissue (*fascia*) which imbeds nerves and blood vessels. The sheaths of this connective tissue influence the mechanical properties of muscle; at the ends of the muscle the tissues combine to form tendons which attach the muscle to bones. By weight, muscle consists of 75% water, 20% protein, and 5% other constituents, such as fats, glucose and glycogen, pigments, enzymes, and salts.

The structural elements of skeletal muscles are listed in Table 3-1. Thousands of individual muscle *fibers* run, more or less parallel, the length of the muscle. They are enveloped by a membrane (*sarcolemma*). Inside, they contain sarcoplasm and up to several hundred nuclei, particularly mitochondria.

In the light microscope, muscle fibers appear striped (*striated*): thin and thick, light and dark bands run across the fiber in regular patterns, which repeat each other along the length of the fiber. One such thin dark stripe appears to penetrate the fiber like a thick membrane or disc: this is the so-called z-line, which is of particular interest as discussed later. The distance between two adjacent z-lines (approximately 250 Å) defines the sarcomere.

Table 3-1. Approximate dimensions of muscle components.

Muscle Components	Diameter	Length
Fiber	$5 * 10^5$ to 10^6 Å	up to 50 cm
Myofibril contains 400-2500 filaments	10^4 to $5 * 10^4$ Å	
Myofilaments:		
Actin	50 to 70 Å	10^4 Å
Myosin	100 to 150 Å	$2 * 10^4$ Å

$$1 \text{ Å} = 10^{-10} \text{m}$$
$$1 \text{ micron} = 10^{-6} \text{m}$$

Components of Muscle Fiber

Within each muscle fiber, thread-like *fibrils* (also called myofibrils from the Greek *mys*, muscle) each wrapped in a membrane (*endomysium*) are arranged by the hundreds or thousands in parallel. Each of these, in turn, consists of bundles of *myofilaments*, protein rods that lie parallel to each other. Of these, there are two types: myosin and actin filaments. Both are elongated polymerized protein molecules. They have the ability to slide along each other, which is the source of muscular contraction. "Stacks" of alternating myosin and actin rods give the appearance of the stripes and bands crossing the muscle fiber, with the myosin filaments bridging the gaps between the ends of adjacent actin filaments as shown in Figure 3-1.

Each myofibril contains between 100 − 2500 myosin filaments, lying side by side, and about twice this many actin filaments. Small projects from the myosin filaments (looking like miniature golf clubs), called crossbridges, protrude towards neighboring actin filaments. The actin filaments are twisted double-stranded protein molecules, wrapped in a double helix around the myosin molecules. In cross-section, each myosin rod is surrounded by six actin rods in a regular hexagonal array. This is the contracting microstructure, the elastic element, of the muscle.

Between the myofibrils lie a large number of *mitochondria*, elongated cells which are the sites of energy production through ATP-ADP metabolism (discussed in Chapter 8). Spaces between the myofibrils are filled with a network of tubular sacs, channels, and cisterns which is connected with the larger tubular system in the z-disks. This is the *sarcoplasmic reticulum*, the plumbing and fueling system of the muscle. It provides the fluid transport between the cells inside and outside the muscle and also carries chemical and electrical messages. It contains the sarcoplasmic matrix, a fluid which embeds the filaments and contains proteins, glucose and glycogen, fat, phosphate compounds, etc.

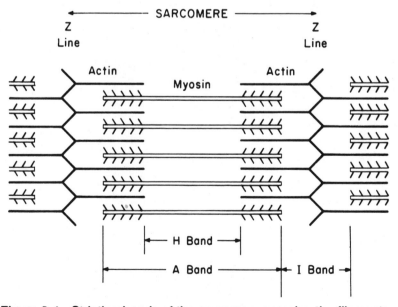

Figure 3-1. Striating bands of the sarcomere crossing the filaments.

Stripes and Bands

The stripes running across the fiber are of varying widths. As Figure 3-1 shows, the space between adjacent ends of actin proteins is called the *H*-band. The length of the myosin rods determine the (anisotropic) *A*-band. The distance between adjacent ends of the myosin rods is the (isotropic) *I*-band. The center of the *I*-band is transversed by the *z*-disk which penetrates the actin myofilaments in their centers and contains the sarcoplasmic reticulum.

Blood Supply to the Muscle

The artery enters the muscle usually at about its half length and branches freely so that final arteries create a complex network infiltrating the muscle tissue. The smallest arteries and their terminal arterioles branch off transversely to the long axes of the muscle fibers, while other arterioles run parallel to individual muscle fibers. Many transverse linkages are present, forming a complex network of blood vessels (the "capillary bed" discussed in Chapter 7). This capillary network is particularly well developed at the motor endplates, discussed later. The abundance of capillaries in the muscle provides good facilities for the supply of oxygen and nutrition to the muscle cells "bathing" in the interstitial fluid (see Chapter 7), and for the removal of metabolic byproducts (see Chapter 8).

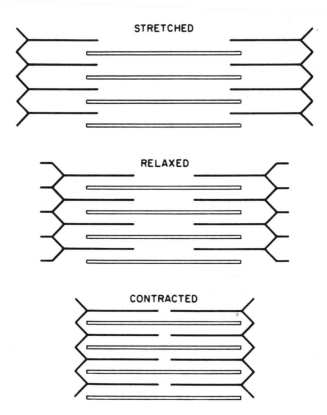

Figure 3-2. Appearance of a sarcomere within stretched, relaxed and contracted muscle.

MUSCLE CONTRACTION

The only active action a muscle can take is to contract; elongation is brought about by external forces that lengthen the muscle. By convention, one distinguishes between the resting length of the muscle, its elongated (stretched) and its contracted length. These three conditions are shown in Figure 3-2. In each case, the width of the A-band remains constant, while the H- and I-bands get narrower in contraction, but wider in stretch as compared to the resting state. Hence, the sarcomere can be lengthened to about 160% of its rest length by stretch or shortened to about 60% in contraction.

Contraction is brought about by the heads of actin rods moving towards each other, which moves the z-line towards the tails of the myosin filaments. This sliding-filament mechanism is caused by mechanical, chemical, and electrostatic forces generated by the interaction of the crossbridges between myosin and actin filaments. (The processes are not yet fully understood.) When the muscle is relaxed, attractive forces between the filaments are neutralized, but after an action potential travels over the muscle fiber membrane, large quantities of calcium ions are released into the sarcoplasm; these ions initiate the contraction force.

Excitation of a muscle follows this sequence:

Step 1. An excitation signal travels along the efferent nervous pathways towards the muscle. The axon of the signal-carrying motor nerve terminates at a "motor endplate" located in a z-disc, which is the myoneural junction for the "motor unit" (see below).

Step 2. The excitation signal (observable in the electromyogram EMG) stimulates a de-polarizing action of the muscle cell membrane. This allows spread of the action potential along the sarcoplasmic reticulum.

Step 3. The potential triggers the release of calcium into the sarcoplasmic matrix surrounding the filaments of the motor unit.

Step 4. This removes the hindrance (*tropomyosin*) for interactions between actin and myosin filaments through chemical, mechanical, and electrostatic actions.

Step 5. Opposing heads of actin rods move towards each other, sliding along the myosin filaments: their heads may meet, even overlap. This reduces the length of the sarcomere, which is called "contraction": the shortening of the myofibrils comprising the muscle fibers under control of the active motor unit. (It should be noted here that — properly used — the term "contraction" only indicates shortening but does not describe the generation of force.)

Step 6. Rebounding of calcium ions in the sarcoplasmic reticulum switches the contraction activity off, allowing the filaments to return to their resting positions.

The "Motor Unit"

Most muscles are joined by hundreds or even thousands of nerve fibers of the efferent (motor) nervous system. The contact between the end point of the axon of one motor-neuron and the sarcolemma of the muscle is called "motor endplate." Each nerve fiber innervates several, usually hundreds to thousands of muscle fibers. These fibers under common control are called a "motor unit": they are all stimulated by the same signal. However, the muscle fibers of one motor unit usually do not lie side by side but are in bundles of only a few fibers each, spread throughout the muscle. Thus, "firing" one motor unit does not cause a strong contraction at one specific location in a muscle but rather a weak contraction throughout the muscle.

Still considering only one motor unit, one can distinguish between the following types of muscular activity.

Twitch. The single contraction resulting from a single instantaneous stimulus followed by complete relaxation. It lasts about 75 to 220 ms, depending on the muscle. A single twitch consists of a latent period, a period of shortening, a period of relaxation, and finally a period of recovery. The latent period, typically lasting about 10 ms, shows no reaction yet of the muscle fiber to the motor neuron stimulus. Shortening takes place usually within 40 ms for a fast-twitch fiber. At the end of this period, the muscle element has reached its shortest length and developed tension. (The energy for this process, mostly generated anaerobically, is freed from the ATP complex — see Chapter 8). The heat energy released causes the crossbridges between actin and myosin to undergo a thermal vibration which results in a kind of "ratchet" action causing the heads of the actin rods to slide towards each other along the myosin filaments. During the relaxation period, also commonly about 40 ms for a fast-twitch fiber, the bridges stop oscillating, the bonds between myosin and actin are broken, and the muscle is pulled back to its original length either by the action of antagonistic muscle or by an external load. (During this period, ATP is resynthesized by ADP. Thus, the primary energy source for muscular contraction is being resupplied.) During the recovery period, again taking about 40 ms, the metabolism of the muscle is aerobic, with glucose and stored glycogen being directly oxydized for the final regeneration of ATP and phosphocreatine. However, if the energy demands on the muscle (through repeated and strong efforts) are beyond the supply capabilities, lactic acid remains as a byproduct of anaerobic glycolysis. If so much lactic acid is built up that the breakdown of ATP becomes blocked, the muscle quickly loses its ability to function and must then rest long enough to deplete both the built-up lactic acid and the oxygen debt incurred (Schneck 1985). Another (supplemental) explanation is the accumulation of potassium (Kahn and Monod 1989).

"Summation" (or "superposition") occurs when twitches are initiated frequently after each other, so that a fiber contraction is not yet completely released by the time the next stimulation signal arrives. In this case, the new contraction builds on a level higher than if the fiber were completely relaxed, and the contraction achieves accordingly higher contractile tension in the muscle. Such "staircase" effect takes place when excitation impulses arrive at frequencies of ten or more per second. When a muscle is stimulated at or above a critical frequency of about 30 to 40 stimuli per second, successive contractions fuse together resulting in a maintained contraction, called tetanus.

Thus, for a single motor unit, the frequency of contractions is controlled by the so-called "rate coding" of the exciting nervous signals. It is of interest to note that with increasing frequency of contraction, building upon each other, the force of contraction also increases. In superposition of twitches, the tension generated may be double or triple as large as in a single twitch and in full tetanization may build up to five times the single twitch tension.

Muscle fatigue. Prolonged and strong contraction of a muscle leads to fatigue, which is the inability of the contractile and metabolic processes to continue the supply of needed energy carriers and to remove metabolic byproducts, particularly lactic acid and potassium. The interruption of blood flow through a muscle (e.g., when a contracted muscle compresses its own blood vessels thus shutting off its own circulation) leads to complete muscle fatigue in about a minute, forcing relaxation. Such fatigue, which may occur (more slowly) even when the muscle is not maximally active, is felt when one works overhead with raised arms, e.g., while fastening a screw in the ceiling of a room. Muscle fatigue in the shoulder muscles makes it impossible to keep one's arms raised after only a minute or so, even though nerve impulses still arrive at the neuromuscular junctions, and the resulting action potentials continue to spread over the muscle fibers.

The higher the requirements are for (isometric) strength exertion of a given muscle, the shorter the period through which this strength can be maintained. Figure 3-3 schematically shows this relation between strength exertion and endurance. Maximal strength can be maintained for only a few (less than 10) seconds; 50% of strength is available for about one minute; but less than 20% can be applied for longer times.

Muscular fatigue can be completely overcome by rest, thus physiological "fatigue" may be defined operationally (in contrast to psychological "boredom") as a "state of reduced physical ability which can be restored by rest."

Muscular Activities

In contrast to the foregoing discussion of a single motor unit, the following text relates to the activities of a whole muscle, comprising several or many motor units. Its contraction activity is controlled by "recruitment coding," that is by how many and which motor units are activated at any given instant. Each single motor unit is triggered to contract, and the cooperative effort of the participating motor units determines the contraction of the whole muscle.

Muscle control depends on the number of muscle fibers innervated by one nerve axon: the larger the ratio, the finer the muscle control. For example, in eye musculature one nerve controls seven muscle fibers, for an innervation ratio of 1:7, while the quadriceps femoris extending the knee has a ratio of approximately 1:1000. Gradation of contraction is also controlled by the portion of total fibers in a muscle excited simultaneously. In general, one cannot voluntarily contract more than two-thirds of all fibers of a muscle at once. But contraction of all fibers at the same time can occur as a result of a proprioceptor reflex; this may strain the muscle to its total structural tensile capacity, and might even tear it (Stegemann 1981).

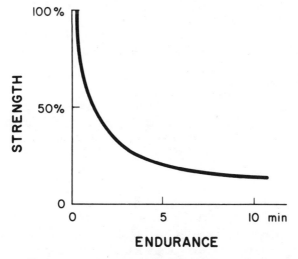

Figure 3-3. Muscle strength and endurance.

The action potential spreads along the muscle at speeds of approximately 1 to 5 ms^{-1}. Muscle fibers may consist of a slow-twitch type (also called Type I or red fiber) which has a relatively slow contracting time (80 to 100 ms until peak is reached) and is well suited for prolonged work because it can work anaerobically. Also, it is triggered by relatively low-rate signals, and therefore often called a low-threshold fiber. In contrast, fast-twitch fibers (also called Type II or white or high-threshold fibers) have relatively short contraction times (about 40 ms) but fatigue rapidly. (Type II-fibers are subdivided again into groups a and b.) The proportion of fiber types in a given individual seems to be genetically determined, but the total number of fibers may be changeable by training (Astrand and Rodahl 1977) — perhaps fibers of one type may be modified into another by use (Astrand and Rodahl 1986). Each motor unit appears to contain only one type of fiber. Slow fibers are mostly recruited for finely controlled actions, while strong efforts generally involve fast fibers. Both fiber types develop about the same tension. A note of caution: Physiological, biomechanical, and morphological properties of muscle fibers vary over a wide continuum of characteristics. Only as one compares far apart segments with each other, one sees distinctions. Furthermore, most research on muscle fiber types has been performed on cats, guinea pigs, and rabbits. In these animals, the muscle fiber distinctions are more clearly apparent than in humans (Basmajian and DeLuca 1985).

Agonist-Antagonist

In the human body, muscles are arranged in groups that oppose each other in their actions around their spanned articulation. One muscle, or one group of synergistic muscles, flexes, while the other extends, as shown in Figure 3-4. This often simultaneous action of agonist (also called protagonist) and antagonist allows the body to control strength and motion.

THE "ALL-OR-NONE PRINCIPLE"

It has been customary to say that all muscle fibers innervated by one motor nerve are either fully relaxed or fully contracted. While this is obviously true for the fully excited or fully relaxed states, it does not apply when a muscle is in the initial phase of twitch buildup or during its return to the resting state after attaining contraction. Furthermore, a muscle may be under tension because of external stretch. Hence, the "all-or-none principle" should not be taken to describe truly all conditions of the muscle.

MUSCLE STRENGTH

The term "voluntary muscle strength," unfortunately often used in a confusing manner, is defined and used here as follows: it "is the torque that a given muscle (or groups of muscles) can maximally develop voluntarily around a skeletal articulation which is spanned by the muscle(s)." For convenience, this torque may be measured as the force acting perpendicularly to a (known) lever arm around this articulation. (For more detail, see Chapter 5.)

This definition acknowledges the fact that it is (with current technology) impossible to measure the force or tension developed within a muscle of a living human. If it becomes feasible to measure this internal force directly, the definition above could be abolished in favor of "the maximal voluntary force that a muscle can exert along its length, under voluntary control."

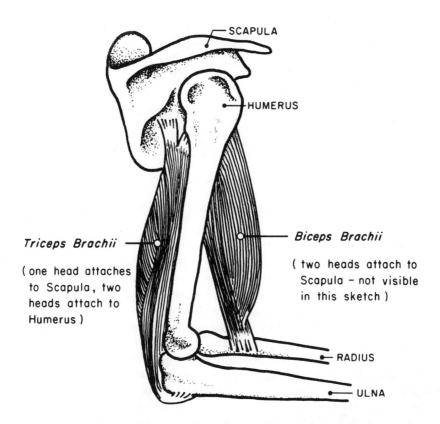

SCAPULA

HUMERUS

Triceps Brachii

(one head attaches
to Scapula, two
heads attach to
Humerus)

Biceps Brachii

(two heads attach to
Scapula – not visible
in this sketch)

RADIUS

ULNA

Figure 3-4. Biceps and triceps muscles as antagonistic pair controlling elbow flexion and extension. Not shown are the brachialis muscle (attaching to humerus and ulna) and the brachioradilis muscle (connecting humerus and radius) which act together with the biceps as a synergistic flexor group.

Measurements on animal muscles have shown that muscle strength depends on muscle length, among other variables. Figure 3-5 shows that the maximal active torque that a muscle can develop is at its resting length. (This "resting" length depends on the lengths of the sarcomeres along the muscle; it is at equilibrium length of the unstimulated muscle detached from its bone or slightly stretched when attached by its tendons to bones — Astrand and Rodahl 1986.) As the muscle shortens by active contraction, the strength falls and reaches zero at its most contracted length, approximately 60% of resting length. Similarly, the active contraction strength is being reduced as the muscle is lengthened beyond its resting length by an external force. However, lengthening of the muscle is accompanied by increasing elastic resistance of the tissue against stretch. Adding this passive resisting strength to the actively developed torque results in the total muscular force between resting and breaking length of the muscle.

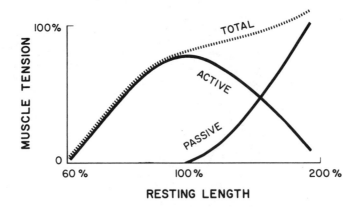

Figure 3-5. Tension within a muscle at different lengths (schematically).

Regulation of Strength Exertion

One way to understand the events involved in the exertion of muscular strength is to use the model in Figure 3-6. The control initiatives generated in the central nervous system start with calling up an "executive program" which, whether innate or learned, exists for all normal muscular activities, such as walking or pushing and pulling objects. The general program is modified by "subroutines" appropriate for the specific case, such as walking quickly upstairs, or pulling hard, or pushing carefully. These in turn are modified by "motivation," which determines how (and how much of the structurally possible) strength will be exerted under the given conditions (Schneck 1985). A qualitative listing of circumstances which may increase or decrease one's willingness to exert strength is given in Table 3-2.

The result of these complex interactions manifests itself in the excitation signals E transmitted along the efferent nervous pathways to the motor units involved where they trigger muscle contractions.

The contraction strength developed in the muscle depends on the motor units involved, on the rate and frequency of signals received, and possibly on existing fatigue in the muscle left from previous contractions.

The output of the muscular contraction is modified by the existing mechanical conditions, such as the lever arms at which muscle tendons pull with respect to the bridged articulation and by the pull angle with respect to the lever arm. These conditions would change, of course, in dynamic activities while they are assumed constant in a static effort.

Thus, the output of this complicated chain of controllers, feedforward signals, controlled elements, and modifying conditions is the "strength" measured at the interface between the body segment involved and the measuring device (the object against which strength is exerted). Of course, the assumption is that at any moment at least as much resistance is available at this interface as can be exerted by a person. If this were not the case, i.e., if Newton's Third Law would be violated, no reliable strength measurement could be performed.

The model also shows a number of feedback loops through which the muscular exertion is monitored for control and modification. The first feedback loop, $F1$, is in fact a reflex-like arc which originates at proprioceptors, such as Ruffini organs in the joints

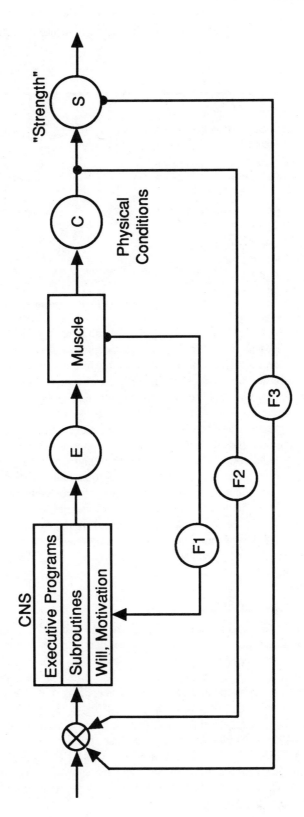

Figure 3-6. Model of the regulation of muscle strength exertion (modified from Kroemer 1979).

E: Efferent excitation impulses generated in the CNS

F: Afferent feedback loops.

Table 3-2. Factors likely to increase (+) or decrease (−) maximal muscular performance.

	Likely Effect
Feedback of results to subject	+
Instructions on how to exert strength	+
Arousal of ego involvement, aspiration	+
Pharmaceutical agents (drugs)	+
Startling noise, subject's outcry	+
Hypnosis	+
Setting of goals, incentives	+ or −
Competition, contest	+ or −
Verbal encouragement	+ or −
Fear of injuries	−
Spectators	?
Deception	?

(signaling location), Golgi tendon-end organs (indicating changes in muscle tension), and muscle spindles (indicating length). These interoceptors and their signals are not under voluntary control and influence the signal generator in the spinal cord very quickly (but how they do it is not well understood). The other two feedback loops originate at exteroceptors and are rooted through a comparator modifying the input signal into the central nervous system. The second loop, $F2$, originates at receptors signaling events related to touch, pressure, and body position in general (kinesthetic signals), such as one feels when pulling on a handle: body position is monitored together with the sensations of pressure in the hand coupled with the instrument, and the sensation of pressure felt at the feet through which the chain of force vectors is transmitted when standing. The third feedback loop, $F3$, originates at exteroceptors and signals such events as sounds and vision related to the effort to the comparator. For example, this may be the sounds or movements generated in the experimental equipment by the exertion of strength; it may be the pointer of an instrument that indicates the strength applied; or it may be the experimenter or coach giving feedback and exhortation to the subject, depending on the status of the effort.

Measurement Opportunities

Which measurement opportunities does this model indicate? Considering the feed-forward section of the model, it becomes apparent that there is (with current technology) no suitable means to measure the executive programs, the subroutines, or the effects of will or motivation on the signals generated in the central nervous system. Only very general information can be gleaned from electroencephalograms (EEGs). Better observable events are the efferent excitation impulses that travel along the motor nerves to the muscles, where they spread from the motor endplates. These signals can be recorded through (intrusive or surface) electrodes that monitor the electric events by electromyograms (EMGs). The contraction activities at the muscle can be observed qualitatively, but not quantitatively in the living human. As mentioned earlier, at this time no instruments are available to measure directly the tensions within muscle filaments, fibrils, fibers, or groups of muscles in situ.

The physical conditions accompanying the strength exertion can be observed and recorded if they are external to the body: location of the coupling between body and measuring device, direction of force or torque, time history of exertion, positioning and support of the body, temperature, humidity, etc. However, the mechanical advantages internal to the body are much more difficult and often practically impossible to record and control. This concerns, for example, lever arms of the tendon attachments and pull angles within the muscle with respect to their lever arm attachments, etc.

Hence, the resulting output, labeled "strength," of this complex system is the only clearly definable and measurable event. The amount and direction of torque or force exerted to and detected by the measuring device, over time, are easily identifiable.

The foregoing remarks indicate the problems and possibilities associated with the measurement of human muscular strength. Apparently, the result of each signal or element in the feedforward loop of the model is modified by following (unknown) factors until finally "strength" is exerted. The effects of muscular contractions are modified by physical conditions, such as mechanical advantages; the consequences of excitation impulses observable in the EMG depend on the conditions of the muscle; the components of the "decision making" in the CNS are not discernible from the EEG. Hence, with current techniques, the only unambiguous measurement location is at the instrument that records the combined output of all components in the loop: strength is identified by what is measured externally.

The feedback loops in the nervous system offer some interesting possibilities. Unfortunately, the afferent pathways from interoceptors are anatomically and functionally associated with the feedforward lines for the efferent impulses. Hence, it is (with current technology) practically impossible to distinguish the electric events associated with the feedback signals from those associated with the feedforward signals. Measurement and control of the first two feedback loops is not easily accomplished, currently in fact not practicable for strength measurements. The third feedback loop, starting at exteroceptors, such as in the eyes and ears can be experimentally manipulated (Kroemer and Marras 1980). This allows some experimental control over the subjects but does not provide a means to measure strength.

Hence, the following two conclusions are obvious for strength measurements on living humans:

1. Muscular strength is what is measured by an instrument.
2. Strength is influenced by motivation and the physical conditions under which it is exerted.

ASSESSMENT OF HUMAN MUSCLE STRENGTH

Generation of muscle strength is a complex procedure of myofilament activation through nervous feedforward and feedback control, and the use of mechanical leverages within the body. Mechanically, the main distinction between muscle activation has been between "statics" and "dynamics." In physiological terms, the static condition is generated in an *isometric* muscle contraction where, presumably, (after an initial shortening) the muscle length remains constant. (The Greek term *iso* means unchanged or constant, and *metrein* refers to the measure or length of the muscle.) If there is no change in muscle length during the isometric effort, there is no motion of the involved body segments, hence Newton's First Law requires that all forces acting within the system are in equilibrium. Because no displacement results from the muscular contraction, the physiological "isometric" case is equivalent to the "static" condition in physics.

This theoretically simple and experimentally well controllable condition has lent itself to rather easy measurement of muscular strength, and most of the information currently available on "human strength" is limited to the static (isometric) muscular effort.

Currently no technology exists to measure a muscular effort directly at the human muscle *in vivo et situ*: there are no instruments that can be inserted right at the muscle. Instead, measurements of muscular efforts are done at the outside of the body, where it interfaces with some kind of a dynamometer or other measuring instrument. Nevertheless, the measurements taken at the outside of the body are taken as indicative of the events taking place at the muscle. For isometric (static) efforts, the measurement techniques have been traditionally confined to either taking the "peak" strength observed during the effort as the indicator of muscle strength; an alternative procedure is to take an average measure over the exertion period of a few seconds (Caldwell, Chaffin, Dukes-Dobos, Kroemer, Laubach, Snook, and Wasserman 1974).

Dynamic muscular efforts are much more difficult to describe and control than static contractions. In dynamic activities, muscle length changes, and therefore involved body segments move. Thus, displacement is present, and its time derivatives (velocity, acceleration, and jerk) must be considered. This is a much more complex task for the experimenter than encountered in static testing. Only recently, a systematic breakdown into independent and dependent experimental variables has been presented for dynamic and static efforts (Kroemer, Marras, McGlothlin, McIntyre, and Nordin 1989). *Independent* variables are those that are purposely manipulated during the experiment in order to assess resulting changes in the *dependent* variables. For example, if one sets the displacement (muscle length change) to zero — the isometric condition — one may measure the force generated and possibly the number of repetitions that can be performed until force is reduced because of muscular fatigue. This case is described in Table 3-3. Of course, with no displacement, its time derivatives velocity, acceleration, and jerk are also zero. In the isometric technique, one is also likely to control the mass properties, probably by keeping them constant.

One may also chose to control velocity as an independent variable, i.e., the rate at which muscle length changes. If velocity is set to a constant value, one speaks of *isokinetic* muscle strength measurement. Time derivatives of constant velocity, acceleration and jerk, are zero. Mass properties are usually controlled in isokinetic tests. The variables displacement, force, and repetition can either be chosen as dependent variables or they may be controlled independent variables. Most likely, force and/or repetition are the dependent variables chosen to assess the result of the testing. Following the scheme laid out in Table 3-3, one also can devise tests in which acceleration or its time derivative, jerk, are kept constant. While these test conditions are theoretically possible, they have so far not been commonly applied.

If one sets the amount of force (or torque) to a constant value, it is most likely that mass properties and displacement (and its time derivatives) are controlled independent variables, and repetition a dependent variable. This *isoforce* condition in which muscle tension is kept constant (*isotonic*) is, for practical reasons, often combined with an isometric condition, such as in holding a load motionless (i.e., displacement is zero).

Note that the term *isotonic* has often been wrongly applied. Some older textbooks used lifting or lowering of a constant mass (weight) as the typical example for isotonics. This is physically false for two reasons. The first is that according to Newton's Laws accelerations and decelerations of a mass requires changing (not: constant) forces that need to be applied. The second fault lies in overlooking the changes that occur in the mechanical conditions (pull angles and lever arms) under which the muscle functions during the

Table 3-3. Techniques to measure motor performance by selecting specific independent and dependent variables.

Names of Technique / Variables	Isometric (Static) Indep.	Dep.	Isokinetic Indep.	Dep.	Isoacceleration Indep.	Dep.	Isojerk Indep.	Dep.	Isoforce Indep.	Dep.	Isoinertial Indep.	Dep.	Free Dynamic Indep.	Dep.
Displacement, linear/angular	constant* (zero)		C	or X	C	or X	C	or X	C	or X	C	or X		X
Velocity, linear/angular	O		constant		C	or X	C	or X	C	or X	C	or X		X
Acceleration, linear/angular	O		O		constant		C	or X	C	or X	C	or X		X
Jerk, linear/angular	O		O		O		constant		C	or X	C	or X		X
Force, Torque	C	or X	C	or X	C	or X	C	or X	constant		C	or X		X
Mass, Moment of Inertia	C		C		C		C		C		constant		C	or X
Repetition	C	or X	C	or X	C	or X	C	or X	C	or X	C	or X	C	or X

Legend
C = variable can be controlled
* = set to zero
O = variable is not present (zero)
X = can be dependent variable

The boxed constant variable provides the descriptive name.

activity. (See Chapter 5 for more detail.) Hence, even if there were a constant force to be applied to the external object (which is not the case), the changes in mechanical advantages would result in changes in muscle tonus. It is certainly misleading to label all dynamic activities of muscles isotonic, as is unfortunately done occasionally.

In the *isoinertial* condition, the mass properties are controlled, usually set to a constant value. In this case, repetition of moving such constant mass (as in lifting) may either be a controlled independent or more likely be a dependent variable. Also, displacement and its derivatives may become dependent outputs. Force (or torque) applied is likely to be a dependent value, according to Newton's Second Law (*force* equals *mass* times *acceleration*).

Table 3-3 also contains the most general case of motor performance measurement, labeled *free dynamic*. In this case, the independent variables displacement and its time derivatives, as well as force are unregulated, i.e., left to the free choice of the subject. Only mass and repetition are usually controlled but may be used as dependent variables. Displacement and its time derivatives may be dependent variables. Force, torque, or some other performance measure is likely to be chosen as a dependent output.

The foregoing discussion of dynamic measurement indicates that these activities are much more complex to describe and control than the static (isometric) contraction. Dynamics involve motions with changes in muscle lengths. Such displacements can occur over various distances and during variable time periods. This complexity explains why, in the past, dynamic measurements other than isokinetic and isoinertial testing have been rarely performed in the laboratory. On the other hand, if one is free to perform as one pleases, such as in the "free dynamic" test, very little experimental control can be executed.

Strength Measurement Devices

Devices to measure force or torque ("strength") consist of several components. The first is the *sensor*, the element that experiences the strain generated by force or torque application, and the second the *converter*, the element which changes the strain into a measurable output. Both elements together are often called the *transducer*. A typical transducer consists of a deformable object, such as a metal beam, which is bent (usually only imperceptably little) under the strength exerted on it. A strain gauge is commonly employed to convert the deformation into an electrical signal that is analog to the strain and deformation. (In this context it is rather interesting to note that force and torque are not basic units in the international measurement system but derived from the product of mass and acceleration.) There are no instruments available at this time which are directly sensitive to force or torque; all rely on the sensor-converter technique.

The output of the transducer is usually fed through an *amplifier* so that the signal can be easily transmitted and used. The next element in the measurement device is an *indicator* (display). This is typically a pointer displaced from its zero position according to the strength of the signal received from the amplifier; it may move over a stationary scale, or leave a permanent mark on a strip chart. The displacement of these indicators is analog to the signal. In contrast, other indicators are digital, giving the signal in discrete numbers. Usually, the system also includes a *recorder*, which may be in series or parallel with the indicator. (In some cases an indicator is not used in the system with a recorder.) This serves to record the signals received from the amplifier so that an *analysis* can be performed on the data. The analysis is strictly not part of the measurement chain. However, selection of the statistical analysis technique may have determining effects on the selection of the measurement device itself.

The functions of output amplification, indication, and storage are often combined in computer systems, which may also be programmed to perform the data analysis.

One important aspect is unfortunately too often overlooked: whether home-built or purchased, the measurement device must be calibrated so that it is assured that the same input results in the same known output in each test. It is discouraging to see in a laboratory a measurement device that has not been checked and calibrated for long periods of time.

The Strength Test Protocol

After choosing the type of strength test to be done, and the measurement techniques and the measurement devices, an experimental protocol must be devised. This includes the selection of subjects, their information and protection; the control of the experimental conditions; the use, calibration, and maintenance of the measurement devices; and (usually) the avoidance of training and fatigue effects. Regarding the selection of subjects, care must be taken that the subjects participating in the tests are in fact a representative sample of the population for which data are to be gathered. Regarding the management of the experimental conditions, the control over motivational aspects is particularly difficult. It is widely accepted (outside sports and medical function testing) that the experimenter should not give exhortations and encouragements to the subject (Caldwell, Chaffin, Dukes-Dobos, et al. 1974). The so-called "Caldwell Regimen" pertains primarily to isometric strength testing but partly (indicated by an asterisk *) also applies to dynamic tests. Excerpts read as follows:

Definition: Static strength is the capacity to produce torque or force by a maximal voluntary isometric muscular exertion. Strength has vector qualities and therefore should be described by magnitude and direction.

1. Static strength is measured according to the following conditions.
    (a) Static strength is assessed during a steady exertion sustained for four seconds.
    (b) The transient periods of about one second each, before and after the steady exertion, are disregarded.
    (c) The strength datum is the mean score recorded during the first three seconds of the steady exertion.
2. (a)* The subject should be informed about the test purpose and procedures.
    (b)* Instructions to the subject should be kept factual and not include emotional appeals.
    (c) The subject should be instructed to "increase to maximal exertion (without jerk) in about one second and maintain this effort during a four second count."
    (d)* Inform the subject during the test session about his/her general performance in qualitative, non-comparative, positive terms. Do not give instantaneous feedback during the exertion.
    (e)* Rewards, goal setting, competition, spectators, fear, noise, etc., can affect the subject's motivation and performance and, therefore, should be avoided.
3.* The minimal rest period between related efforts should be two minutes; more if symptoms of fatigue are apparent.
4.* Describe the conditions existing during strength testing:
    (a) Body parts and muscles chiefly used.
    (b) Body position.

        (c)* Body support/reaction force available.

        (d)* Coupling of the subject to the measuring device (to describe location of the strength vector).

        (e)* Strength measuring and recording device.

5.   Subject description:

        (a)* Population and sample selection.

        (b)* Current health and status: medical examination and questionnaire are recommended.

        (c)* Gender.

        (d)* Age.

        (e)* Anthropometry (at least height and weight).

        (f)* Training related to the strength testing.

6.   Data reporting:

        (a)* Mean (median, mode).

        (b)* Standard deviation.

        (c)* Skewness.

        (d)* Minimum and maximum values.

        (e)* Sample size.

SUMMARY

- Muscle contraction is brought about by shortening of muscle substructures. Elongation of the muscle is due to external forces.

- Muscle contraction is controlled by excitation signals from the Central Nervous System. Each specific signal affects those fibers that are combined to a "motor unit," of which there are many in a muscle.

- An efferent stimulus arriving from the CNS brings about a "single twitch" contraction of the motor unit. A rapid sequence of stimuli can lead to a superposition of muscle twitches which may fuse together into a sustained contraction called tetanus.

- Prolonged and strong contraction leads to muscular fatigue, which hinders the continuation of metabolic processes by reducing oxygen supply to and metabolite removal from the muscle. Hence, maximal voluntary contraction can be maintained for only a few seconds.

- In isometric contraction, muscle length remains constant, which establishes a static condition for the body segments affected by the muscle. In an isotonic effort, the muscle tension remains constant, which usually coincides with a static (isometric) effort.

- Dynamic activities result from changes in muscle length, which bring about motion of body segments. In an isokinetic effort, speed remains unchanged. In an isoinertial test, the mass properties remain constant.

- Muscle strength is the maximal torque that can be developed voluntarily by muscle(s) around a body articulation. Thus, by definition, a maximal voluntary effort depends on the motivation of the person exerting the effort.

- In the living human, muscle strength is measured as the torque (or force at a given lever arm) exerted to an instrument external to the body.

- Measurement of muscle strength requires carefully controlled experimental conditions.

REFERENCES

Astrand, P. 0. and Rodahl, K. 1977 and 1986. *Textbook of Work Physiology*. Second and Third Edition. New York, NY: McGraw-Hill.

Basmajian, J. V. and DeLuca, C. J. 1985. *Muscles Alive*. Fifth Edition. Baltimore, MD: Williams and Wilkins.

Caldwell, L. S., Chaffin, D. B., Dukes-Dobos, F. N., Kroemer, K. H. E., Laubach, L. L., Snook, S. H., and Wasserman, D. E. 1974. A Proposed Standard Procedure for Static Muscle Strength Testing. *American Industrial Hygiene Association Journal* 35(4):201–206.

Kahn, J. F. and Monod, H. 1989. Fatigue Induced by Static Work. *Ergonomics*, 32:7:839–846.

Kroemer, K. H. E. 1979. A New Model of Muscle Strength Regulation. In *Proceedings, Annual Conference of the Human Factors Society*, Boston, MA (pp. 19–20). Santa Monica, CA: Human Factors Society.

Kroemer, K. H. E. and Marras, W. S. 1980. Toward an Objective Assessment of the Maximal Voluntary Contraction Component in Routine Muscle Strength Measurements. *European Journal of Applied Physiology*, 45:1–9.

Kroemer, K. H. E., Marras, W. S., McGlothlin, J. D., McIntyre, D. R., and Nordin, M. 1989. Assessing Human Dynamic Muscle Strength. (Technical Report, 8-30-89). Blacksburg, VA: Virginia Tech, Industrial Ergonomics Laboratory.

Schneck, D. J. 1985. Deductive Physiologic Analysis in the Presence of "Will" as an Undefined Variable. *Mathematical Modelling*, 2:191–199.

Schneck, D. J. 1985. *Biomechanics of Striated Skeletal Muscle*. Santa Barbara, CA: Kinko.

Stegemann, J. 1981. *Exercise Physiology*. Chicago, IL: Yearbook Medical Publishers.

FURTHER READING

Astrand, P. 0. and Rodahl, K. 1986. *Textbook of Work Physiology*. Third Edition. New York, NY: McGraw-Hill.

Basmajian, J. V. and DeLuca, C. J. 1985. *Muscles Alive*. Fifth Edition. Baltimore, MD: Williams and Wilkins.

Chaffin, D. B. and Andersson, G. B. J. 1984. *Occupational Biomechanics*. New York, NY: Wiley.

Daniels, L. and Worthingham, C. 1980. *Muscle Testing*. Fourth Edition. Philadelphia, PA: Saunders.

Guyton, A. C. 1979. *Physiology of the Human Body*. Fifth Edition. Philadelphia, PA: Saunders.

Ingels, N. B. 1979. *Molecular Basis of Force Development in Muscle*. Palo Alto, CA: Palo Alto Medical Research Foundation.

Kroemer K. H. E. 1970. Human Strength: Terminology, Measurement and Interpretation of Data. *Human Factors*, 12:279–313.

Phillips, C. A. and Petrofsky, J. S. (eds). 1983. *Mechanics of Skeletal and Cardiac Muscle*. Springfield, IL: Thomas.

Stegemann, J. 1981. *Exercise Physiology*. Chicago, IL: Yearbook Medical Publishers.

Weller, H. and Wiley, R. L. 1979. *Basic Human Physiology*. New York, NY: Van Nostrand.

APPENDIX: GLOSSARY OF MUSCLE TERMS

acceleration — Second time derivative of displacement.

action, activation (of muscle) — See contraction.

concentric (muscle effort) — Shortening of a muscle against a resistance.

contraction — Literally, "pulling together" the Z lines delineating the length of a sarcomere, caused by the sliding action of actin and myosin filaments. Contraction develops muscle tension only if the shortening is resisted.

Note that during an isometric "contraction" no change in sarcomere length occurs and that in an eccentric "contraction" the sarcome is actually lengthened. To avoid such contradiction in terms, it is often better to use the terms activation effort, or exertion.

displacement — Distance moved (in a given time).

distal — Away from the center of the body.

dynamics — A subdivision of mechanics that deals with forces and bodies in motion.

eccentric (muscle effort) — Lengthening of a resisting muscle by external force.

effort (of muscle) — See contraction.

exertion (of muscle) — See contraction.

fibers — See muscle.

fibrils — See muscle.

filaments — See muscle.

free dynamic — In this context, an experimental condition in which neither displacement and its time derivatives, nor force are manipulated as independent variables.

isoacceleration — A condition in which the acceleration is kept constant.

isoforce — A condition in which the muscular force (tension) is constant. This term is equivalent to isotonic.

isoinertial — A condition in which muscle moves a constant mass.

isojerk	A condition in which the time derivative of acceleration, jerk, is kept constant.
isokinetic	A condition in which the velocity of muscle shortening (or lengthening) is constant. (Depending on the given biomechanical conditions, this may not coincide with a constant angular speed of a body segment about its articulation.)
isometric	A condition in which the length of the muscle remains constant.
isotonic	A condition in which muscle tension (force) is kept constant — see isoforce. (In the past, this term was occasionally falsely applied to any condition other than isometric.)
jerk	Third time derivative of displacement.
kinematics	A subdivision of dynamics that deals with the motions of bodies, but not the causing forces.
kinetics	A subdivision of dynamics that deals with forces applied to masses.
mechanical advantage	In this context, the lever arms (moment arms, leverages) at which a muscle works around a bony articulation.
mechanics	the branch of physics that deals with forces applied to bodies and their ensuing motions.
motor unit	All muscle filaments under the control of one efferent nerve axon.
muscle	A bundle of fibers, able to contract or be lengthened. Specifically, striated muscle (skeletal muscle) that moves body segments about each other under voluntary control.
muscle contraction	The result of contractions of motor units distributed through a muscle so that the muscle length shortens.
muscle fibers	Elements of muscle, containing fibrils.
muscle fibrils	Elements of muscle fibrils, containing filaments.
muscle filaments	Muscle fibril elements (polymerized protein molecules) capable of sliding along each other, thus shortening the muscle and, if doing so against resistance, generating tension.
myo	A prefix referring to muscle.

proximal	Towards the center of the body.
rate coding	The time sequence in which efferent signals arrive at a motor unit and cause contractions.
recruitment coding	The time sequence in which efferent signals arrive at different motor units and cause them to contract.
repetition	Performing the same activity more than once. (One repetition indicates two exertions.)
rhythmic	The same action repeated in equal intervals.
statics	A subdivision of mechanics that deals with bodies at rest.
velocity	First time derivative of displacement.

CHAPTER 4

THE NEUROMUSCULAR CONTROL SYSTEM

OVERVIEW

The central nervous system is one of several control and regulation systems of the body. It collects inputs from various sensors that respond to internal and external stimuli. Its integration and regulation functions concerning motor activities are mainly in the cerebrum, the cerebellum, and the spinal cord. The pathways for incoming and outgoing signals are the neurons, which possess the ability to inhibit or facilitate the transmission of impulses.

The Model

The nervous system transmits (feedback) information about events outside and inside the body from various sensors along its afferent pathways to the brain. Here, decisions about appropriate actions and reactions are made, (feedforward) signals are generated and sent along the efferent pathways to the muscles.

INTRODUCTION

The purpose of the human regulatory and control systems is to maintain equilibrium (homeostasis) on the cell level and throughout the body despite changes in the strain on the body generated by varying external environments and work requirements. A detailed consideration of the several parts and various mechanisms of this control system is challenging and stimulating but exceeds the scope of this book. This text primarily considers the neuromuscular system, which is rather well researched.

Regarding the control of muscular action, the human body is under the control of a dual system: both the hormonal (*endocrine*) system and the nervous system have similar functions. Within the hormonal system, one set of hormones (*norepinephrine*) stimulates smooth muscle in some organ, but in others inhibits the muscular contractions; another hormone (*acetylcholine*) has just the opposite effect on the same smooth muscles. Not much is known about the exact locations in the brain that regulate functions of the autonomic nervous system. The following text concentrates on the nervous system, particularly as it affects functions of skeletal muscle.

ORGANIZATION OF THE NERVOUS SYSTEM

Anatomically, one divides the nervous system into three major subdivisions: the *central* nervous system includes brain and spinal cord; it has primarily control functions. The *peripheral* nervous system includes the cranial and spinal nerves; it transmits signals, but usually does not control. The *autonomic* nervous system includes the sympathetic and the parasympathetic subsystems which together regulate, among other involuntary functions, those of smooth and cardiac muscle, of blood vessels, digestion, and glucose release in the liver. The autonomic system generates the "fright, flight or fight" responses.

Functionally, there are two subdivisions of the nervous system: the *autonomic* (visceral) nervous system and the *somatic* nervous system, which controls mental activities, conscious actions, and the skeletal muscles.

Sensors of the Peripheral Nervous System

The Central Nervous System, CNS, receives information concerning the outside from external receptors (*exteroceptors*) which respond to light, sound, touch, temperature, and chemicals; and from internal receptors (*interoceptors*) which report changes within the body. Since all of these sensations come from various parts of the body (Greek, *soma*) external and internal receptors together are also called somesthetic sensors.

Internal receptors include the *proprioceptors*. Among these are the muscle spindles, which are nerve filaments wrapped around small muscle fibers; they detect the amount of stretch of the muscle. Golgi organs are associated with muscle tendons and detect their tension, hence report to the central nervous system information about the strength of contraction of the muscle. Finally, Ruffini organs are kinesthetic receptors located in the capsules of articulations. They respond to the degree of angulation of the joints (joint position), to change in general, and also to the rate of change.

The sensors in the vestibulum are also proprioceptors which detect and report the position of the head in space and respond to sudden changes in its attitude. This is done by sensors in the semi-circular canals, of which there are three, each located in another orthogonal plane. To relate the position of the body to that in the head, proprioceptors in the neck are triggered by displacements between trunk and head.

Another set of interoceptors, called *visceroceptors*, reports on the events within the visceral (internal) structures of the body, such as organs of the abdomen and chest, as well as on events in the head and other deep structures. The usual modalities of visceral sensations are pain, burning sensations, and pressure. Since the same sensations are also provided from external receptors and since the pathways of visceral and external receptors are closely related, information about the body is often integrated with information about the outside.

External receptors provide information about the interaction between the body and the outside: sight (vision), sound (audition), taste (gustation), smell (olfaction), temperature, chemical agents, and touch (taction). Several of these are of particular importance for the control of muscular activities: the sensations of touch, pressure, and pain can be used as feedback to the body regarding the direction and intensity of muscular activities transmitted to an outside object — see Chapter 3. Free nerve endings, Meissner's and Pacinian corpuscles, and other receptors are located throughout the skin of the body, however in different densities. They transmit the sensations of touch, pressure, and pain. Since the nerve pathways from the free endings interconnect extensively, the sensations reported are not always specific for a modality; for example, very hot or cold sensations can be associated with pain, which may also be caused by hard pressure on the skin.

Almost all sensors respond vigorously to a change in the stimulus but will report less and less in the next seconds or minutes if the load stays constant. This adaptation makes it possible to live with, for example, the continued pressure of clothing. The speed of adaptation varies with the sensors; furthermore, the speeds with which the sensations are transmitted to the central nervous system are quite different for different sensors — light and sound, for example, cause the fastest and pain often the slowest reactions.

The Central Nervous System

The brain is usually divided into forebrain, midbrain, and hindbrain. Of particular interest for the neuromuscular control system is the forebrain with the cerebrum, which consists of the two cerebral hemispheres. Voluntary movements are controlled, and sensory experience, abstract thought, memory, learning, and consciousness are located in the cerebrum. The motor cortex controls voluntary movements of the skeletal muscle, while the sensory cortex interprets sensory inputs. The basal ganglia of the midbrain are composed of large pools of neurons, which control semivoluntary complex activities such as walking. Part of the hindbrain is the cerebellum, which distributes and integrates impulses from the cerebral association centers to the motor neurons in the spinal cord.

The spinal cord is an extension of the brain. The uppermost section of the spinal cord contains the twelve pairs of cranial nerves, which serve structures in the head and neck, as well as the lungs, heart, pharynx and larynx, and many abdominal organs. They control eye, tongue, facial movements, and the secretion of tears and saliva. Their main inputs are from the tastebuds in the mouth, the nasal olfactory receptors, and touch, pain, heat, and cold receptors of the head. Thirty-one pairs of spinal nerves pass out between the appropriate vertebrae and serve defined sectors of the body. Nerves are mixed sensory and motor pathways, carrying both somatic and autonomic signals between the spinal cord and the muscles, articulations, skin, and visceral organs.

THE NERVOUS PATHWAYS

Information from a sensor is passed along a feedback (afferent) path to the decision maker in the spinal column or brain and then as a feedforward signal along the efferent pathway to the effecting organ. Paths leading to muscles are also called motor pathways. The basic functional unit of this transmission system is the nerve cell, the neuron. Neurons transmit signals from one to another through their filamentous nerve fibers. In the brain, neurons are also responsible for the storage of memories, for patterns of thinking, for initiating motor responses, etc. There are about twelve billion neurons in the brain and spinal cord. A synapse is the junction between two neurons, through which signals are transmitted. The synapse has a switching ability, that is, it may or may not transmit signals.

The Neuron

Figure 4-1 sketches a "typical" neuron combining the main features of motor, sensory, and interneurons. It consists of three major parts: the main body, the *soma*, and two different processes, branching *dendrites* and a long extension, the *axon*. Each motor neuron has only one axon, and signals are transmitted from the cell body outward through its axon. At a distance from the soma, the axon branches out into terminal fibrils that connect with other neurons. The length of the axon may be only a few millimeters, or a meter or longer. The neuron has up to several hundred dendrites, projections of usually

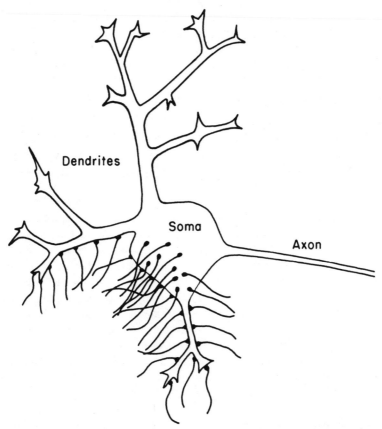

Figure 4-1. Typical neuron components: soma, dendrites, axon. Synapses are sketched only in the low left part of the figure.

only a few millimeters. They receive signals from the axons of other neurons and transmit these into their own neuron cell body. Shown (only in the lower left corner of Figure 4-1) are *synapses*, the endpoints of (hundreds of) fibrils coming from axons of other neurons. Synapses are formed like knobs, bulbs, clubs, or feet. Each synaptic body has numerous vesicles which contain transmitter substance. The synaptic membrane is separated from the opposing subsynaptic membrane of the neuron by a synaptic cleft, a space of about 2 Å.

Transmission of Nerve Signals

Some axon terminals secrete an excitory transmitter substance while others carry an inhibitory neurohumor; this means that some terminals excite the neuron and others inhibit it. Among the excitory neurohumors is norepinephrine, while one of the inhibitory secretions is dopamine. Certain chemical agents (such as some anaesthetics and curare) can prevent the secretion of excitory neurohumor thus inhibiting the transmission of signals, possibly causing paralysis of the musculature.

Excretion of an excitatory transmitter increases the permeability of the subsynaptic membrane beneath the synaptic knob, which allows sodium ions to flow rapidly to the inside of the cell. Since sodium ions carry positive charges, the result is an increase in positive charge inside the cell, bringing about an excitatory post-synaptic potential (see the following discussion of the "sodium pump"). This sets up an electrical current throughout the cell body and its membrane surfaces, including the base of the axon. If this potential becomes high enough, it initiates an "action potential" in the axon. If the potential does not exceed a threshold value, no action impulse will be transmitted along the axon. Usually, the threshold is exceeded by a summation of inputs. Such summation can either result from simultaneous firing of several synapses or from rapid repeated firing of one synapse. If the threshold is exceeded, the axon carries repeated action impulses as long as the potential remains above the threshold. The higher the post-synaptic potential rises, the more rapidly the axon fires. Motor signal intensities are typically within the range of ± 70 mV along the axon.

Depending on their type and size, some neurons are able to transmit as many as a thousand impulses per second along their axons, while others may not be able to transmit more than twenty-five signals per second.

The inhibitory transmitters have an opposite effect on the synapse: they create a negative potential called the inhibitory post-synaptic potential. This reduces the actual synaptic potential, which may result in a value below the threshold of the neuron: different neurons have different thresholds. Thus, the inhibitory synaptic knobs may stop or prevent neuron discharge.

Most nervous impulses are not carried by a single neuron from receptor to destination but follow a chain of linked neurons. The transmission can be interrupted if it reaches a neuron with a particularly high threshold. Once developed, the action potential propagates itself point by point along a nerve fiber without needing further stimulation. At each successive point along the fiber, the action potential rises to a maximum and then rapidly declines. Because of this shape, the action potential is occasionally called spike potential. The height of its spike, its strength, does not vary with the strength of the stimulus, nor does it weaken as it progresses along the axon. Either the stimulus is strong enough to elicit a response, or there is no potential; this is an all-or-none phenomenon similar to the one discussed in the contraction of skeletal muscle.

While nerve fibers fatigue hardly at all, synapses fatigue, some very rapidly, some rather slowly.

The velocity of the nerve impulse is a constant for each nerve fiber, ranging from 0.5 to about 150 ms^{-1}. This speed does not change along a particular fiber and is correlated with the diameter of the fiber, being faster in a thick fiber than in a thin fiber. In the peripheral parts of the nervous systems, axons and dendrites are sheathed along their length. The envelope consists of myelin, a white material composed of protein and phospholipid. The presence of a myelin sheath allows a larger speed of conduction. Skeletal muscles are served by thick myelinated axons terminating at the motor endplates, while pain fibers are the thinnest and non-myelinated.

Nerve fibers are extensions from a single neuron, hence part of a single cell. Nerves, in contrast, are multi-cellular structures, bundles of nerve fibers gathered from many neurons and arranged somewhat like the wires within a cable.

The "Sodium Pump"

In an excited neuron the potential change is brought about by a redistribution of the positive sodium (Na) and potassium (K) ions on either side of the cell membrane. With a change of permeability of the membrane, sodium ions rush inward and potassium ions move out of the fiber. However, the positive Na ions overshoot an equilibrium point and become more concentrated inside the fiber than in the interstitial fluid. This causes the interior of the fiber to become positively charged and the external surface to become "negative," that is, less positive. At this moment, the polarity of the membrane has been established as a change from its state of an undisturbed cell. When the positive K ions move out, the membrane becomes re-polarized, re-establishing a positive charge on its external surface and leaving the cell interior less positive, that is, more negative. When the action potential moves on to the neighboring region on the fiber, this "sodium (and potassium) pump" restores the original ionic balance: the positive Na ions are moved out and the positive K ions are re-captured.

Reflexes

The spinal cord is a center of coordination for certain actions, particularly limb movements, which do not need to involve the brain proper. Such a reflex usually begins with a stimulation of a peripheral sensory receptor. Its signal is sent as afferent impulse to the spinal cord, where it evokes a quick response which is sent as efferent signal to the appropriate muscles. In this way, a reactive action of an effector muscle can be executed a few milliseconds after the stimulus was received since no time-consuming higher brain functions are involved. All effectors are either muscle fibers or gland cells, hence the result of a reflex is either a muscular contraction or a gland secretion.

Control of Muscle Movement

While some motor functions of the body can be performed without involvement of higher brain centers, many complex voluntary muscular activities need fine regulation. These require various degrees of involvement of the higher brain centers, such as the cerebral cortex, the basal ganglia, and the cerebellum. It is believed that the motor cortex controls mostly very fine discrete muscle movements, while the basal ganglia have large pools of neurons organized for the control of complex movements, such as walking, running, and posture control. (This may be the locus for the "executive programs" mentioned in Chapter 3.) The motor pathway is the location at which efferent signals can be picked up by wire, needle, or surface electrodes and recorded by their electrical activities in an electromyogram (EMG). A suitable location for electrode placement is near the motor endplates of the innervated motor unit of the muscle of interest — see Chapter 3.

SUMMARY

- The body must continuously control muscle functions according to information about conditions and events reported from various sensors. The information feedback to the central nervous system (where decisions are made and actions initiated) and the feedforward signals to the muscles flow along the peripheral nervous system.
- Afferent (feedback) and efferent (feedforward) signals as transmitted along neurons, consisting of soma, dendrites, and axon. At the neuron, nerve fibrils (from other neurons) end in synapses which serve as selective switches. Depending on the strength of the incoming signal, it is or is not transmitted across the synaptic membrane.
- With usually many sensors reacting to the same stimulus, and many afferent pathways transmitting the signals at different intensities and speeds, the peripheral nervous system serves as a filter or selector for the central nervous system.

REFERENCES AND FURTHER READING

Astrand, P. 0. and Rodahl, K. 1986. *Textbook of Work Physiology*. Third edition. New York, NY: McGraw-Hill.

Guyton, A. C. 1979. *Physiology of the Human Body*. Fifth edition. Philadelphia, PA: Saunders.

Weller, H. and Wiley, R. L. 1979. *Basic Human Physiology*. New York, NY: Van Nostrand Reinhold.

INTRODUCTION

Biomechanics is not a new science. Leonardo da Vinci (1452–1519) and Giovani Alfonso Borelli (1608–1679) combined physics with anatomy and physiology. In his book Borelli developed *De Motu Animalium* a model of the human skeleton consisting of a series of links (long bones) joined in their articulations and powered by muscles bridging the articulations. This "stick person" approach still underlies many current biomechanical models of the human body.

Development of the biomechanical sciences is closely linked to the physical laws developed by Isaac Newton (1642–1727) and achieved a first high point in the late 1900s when, for example, Harless determined the masses of body segments, Braune and Fischer investigated the interactions between mass distribution and external impulses applied to the human body, and when von Meyer discussed statics and mechanics of the human body. Since then, biomechanical research has addressed, e.g., responses of the human body to vibrations and impacts, human strength and motion regarding the whole body or specific segments thereof, functions of the spinal column, hemodynamics and the cardiovascular system, and prosthetic devices (King and Chou 1976, King 1984).

Despite the large body of information available, biomechanics is still a developing scientific and engineering field with a wide variety of focus points, research methods, and measurement techniques producing new theoretical and practical results (Kroemer, Snook, Meadows, and Deutsch 1988). While using the substantial data and knowledge base existent, the practicing engineer must attentionally follow the progress reported in the scientific and engineering literature in order to stay abreast of the developments so that the newest information can be applied to the design and management of human/ machine systems. (A list of journals in which much biomechanical information appears is at the end of this chapter.)

By treating the human body as a mechanical system, many and gross simplifications are done, for example by disregarding mental functions. Furthermore, many components of the body are simply considered in their mechanical analogies, such as:

Bones: lever arms, central axes, structural members
Articulations: joints and bearing surfaces
Articulation Linings: lubricants, joint structures
Tendons: cables transmitting muscle forces
Tendon Sheaths: pulleys and sliding surfaces
Anthropometric Data: dimensions of body segments, both in their surfaces and internally
Flesh: volumes, masses
Contours: surfaces of geometric bodies
Nerves: control and feedback circuits
Muscles: motors, dampers or locks
Tissue: elastic load bearing surfaces, springs, contours
Organs: generators or consumers of energy

This listing indicates some of the limitations imposed by biomechanical considerations of the human body.

Stress and Strain

In engineering terms, *strain* is the result or effect of *stress*: stress is the input, strain the output. One example: the weight of a truck on a bridge stresses the bridge, generating strain in the bridge structures.

In the 1930s, the psychologist Hans Selye introduced the concept of stress (or distress, if excessive) being caused by *stressors*.

Obviously, the use of the term "stress" as either being the cause or the result can create much confusion. (What is, for example, "job stress"?) To avoid ambiguity, the engineering terminology will be used in this text: stress produces strain.

MECHANICAL BASES

Mechanics is the study of forces and of their effects on masses. *Statics* consider masses at rest or in equilibrium as a result of balanced forces acting on them. In *dynamics*, one studies the motions of masses and their causes — unbalanced external forces.

Dynamics, often called kinesiology when applied to the human body, is again subdivided. *Kinematics* consider the motions (displacements and their time derivatives velocity, acceleration, jerk) but not the forces that bring these about; in contrast, *kinetics* study exactly these forces.

Newton's Laws are basic to biomechanics. The *first law* states that a mass remains at uniform motion (which includes being at rest) until acted upon by unbalanced external forces. The *second law*, derived from the first, indicates that force is proportional to the acceleration of a mass. The *third law* states that action is opposed by reaction.

Newton's second law sheds light on one important factor in biomechanics — force. Force is not a basic unit but a derived one. Relatedly, no device exists that measures force directly. All measuring devices for force (or torque) rely on other physical phenomena which are then transformed and calibrated in units of force (or torque). The events used to assess force are usually either displacement (such as bending of a metal beam) or acceleration experienced by a mass.

The correct unit for force measurement is the *Newton*; one pound-force-unit is approximately 4.45 Newtons, and 1 kg_f (also called 1 kilopond, kp) equals 9.81 N. The pound (lb), ounce (oz), and gram (g) are usually not force but mass units.

Torque (also called moment) is the product of force and its lever arm (distance) to the articulation about which it acts; the direction of the force must be at a right angle to its lever arm. In kinesiology, the lever arm is often called the "mechanical advantage". Force (as well as torque) is a vector, which means that it must be described not only in magnitude but also by direction, by its line of application and even by the point of application.

Figure 5-1 uses the example of elbow flexion to illustrate these relationships. The primary flexing muscle, the biceps brachii, exerts a force M′ at the forearm (radius) with its lever arm *m* around the elbow joint. This generates a torque T.

$$T = m \ M' \qquad (5\text{-}1)$$

Since, by definition, there must be a right angle between lever arm and force vector, the lever arm *m* is smaller both when the arm is highly flexed or extended than when there is a right angle between forearm and upper arm.

Figure 5-2 represents the same condition in a more lifelike and detailed sketch. This shows that the actual direction of the muscle force vector M differs usually from that of

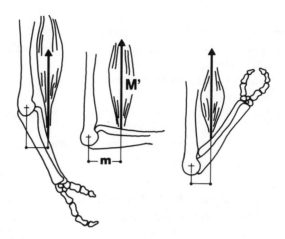

Figure 5-1. Changing lever arm of the force vector M' with varying elbow angle.

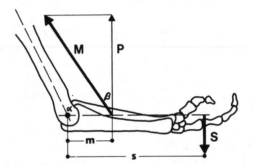

Figure 5-2. Interaction between external force S and muscle force M.

the vector P which is indeed perpendicular to the lever arm m. With the angle β between P and M, the relationship between the two vectors is

$$P = M \cos\beta \qquad (5\text{-}2)$$

The angle β itself varies with the elbow angle α, hence

$$\beta = f(\alpha) \qquad (5\text{-}3)$$

The torque T around the elbow joint generated by the muscle force M is

$$T = m\,P \qquad (5\text{-}4)$$

This torque is counteracted, in the balanced condition, by an equally large but opposing torque which may be generated, for example, by a force vector S pulling down on the hand. With its lever arm s (and assuming that it is perpendicular to its lever arm), S establishes equilibrium if

$$s \ S = m \ P = m \ M \ \cos\beta \qquad (5\text{-}5)$$

Measuring the external force S and knowing the lever arms s and m as well as the angle β, one can solve for the muscle force vector M

$$M = S \ s \ m^{-1} \ (\cos\beta)^{-1} \qquad (5\text{-}6)$$

However, real-life conditions are usually a bit more complex. It is relatively easy to measure the lever arm s of the external force, but it can be rather difficult to determine anatomically the insertion of the tendon connecting the biceps with the radius, establishing the lever arm m of the muscle force vector. Furthermore, the muscle itself is kept partly (by incapsulating ligaments) in a groove of the humerus to near the elbow bend. This makes the angle β steeper near the elbow than assumed in the previous discussions. (Also, the biceps muscle has two heads which split along the upper part of the humerus and attach in different locations to the scapula.) Also, another muscle helps in flexing the elbow: this is the brachialis which originates at about half length of the humerus on its anterior side, and inserts on the ulna; not to mention the brachioradialis, which can also contribute to elbow flexion.

Figure 5-3 illustrates these conditions. M is still the contractile force of the biceps, $(90°\text{-}\beta)$ is its pull angle with respect to the long axis of the forearm, and $P = M \ \cos\beta$ its torque-generating force about the elbow joint at the lever arm m. A similar vector analysis for the force vector N of the brachialis muscle results in its torquing force Q at the lever arm n; γ (also a function of the elbow angle α) is the angle between the directions of N and Q, hence $Q = N \cos\gamma$. Taking into account the external torque $T = s \ S$ as before (assuming a right angle between force S and its lever arm s: if this were not the case, an analysis of S must be done to determine its perpendicular component), the conditions for equilibrium (i.e., no change in elbow angle α as a result of the acting forces) can be stated as follows:

All torques about the elbow joint must sum to zero:

$$m \ P + n \ Q - s \ S = \phi \qquad (5\text{-}7)$$

All forces in vertical directions must sum to zero:

$$P + Q - S = \phi \qquad (5\text{-}8)$$

All forces in horizontal directions must sum to zero:

$$M \ \sin\beta + N \ \sin\gamma + U = \phi \qquad (5\text{-}9)$$

Equation 5-7 indicates that the elbow angle will not change. Equation 5-8 shows that the forearm will not be linearly elevated or lowered. Equation 5-9 demonstrates that there must be a force U — so far not considered — present so that the forearm will not be pushed backward: U may be generated by shoulder muscles acting on the upper arm or by resting the elbow against a vertical stop.

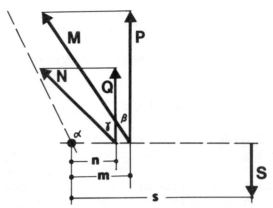

Figure 5-3. Interactions between several muscles and an external force.

This discussion shows that it is relatively easy to measure the resultant output of all concurring muscular forces combined at a suitable interface between body limb and an external measuring device (here represented by force S at lever arm *s*), but the analysis does not indicate which muscles contribute how much. This is illustrated by the fact that the foregoing considerations did not take into account that the triceps muscle may be involved also for control and regulation: It would add to the torque generated by S and counteract the biceps and brachialis muscles, forcing them to increase their efforts. Hence, the analysis technique shows only minimal net results, not the actual efforts of individual muscles which may be much higher than the equations indicate.

ANTHROPOMETRIC INPUTS

Biomechanics rely much on anthropometric data, adapted and often simplified to fit the mechanical approach. In Chapter 1, the reference planes used in anthropometry and subsequently in biomechanics were identified in Figure 1-1: the mid-sagittal plane divides the body into right and left halves; the frontal (or coronal) plane establishes anterior and posterior sections of the body; the transverse (horizontal) plane cuts the body in superior and inferior parts. However, anatomically, only the mid-sagittal plane is well defined in its location, while the frontal and transverse planes need to be fixed by consensus. For this, it is usually assumed that the human stands upright in the so-called anatomical position and that in this case the three planes meet in the center of mass of the body (in the pelvic region), there establishing the origin of an XYZ axis system. Obviously, this convention applies only to the upright standing body. If the posture is different, the location of the center of mass changes (see NASA/Webb 1978; Roebuck, Kroemer, and Thomson 1975; Hay 1973). In this case, one may decide to either retain the anatomical fixation of the coordinate system in the pelvic area or to establish a new origin depending on the given conditions.

For ease of treatment and computation, the human skeletal system is often simplified into a relatively small number of straight-line links (representing long bones) and joints (representing major articulations). Figure 5-4 shows such a typical link-joint system. In this example hands and feet are not subdivided into their components, and the spinal column is represented by only three links. Clearly, such simplification does not represent the true design of the human body but may be sufficient to represent certain mechanical properties.

The determination of the location of the joint center-of-rotation is relatively easy for simple articulations, such as the hinge-types in fingers, elbows, and knees. However, this

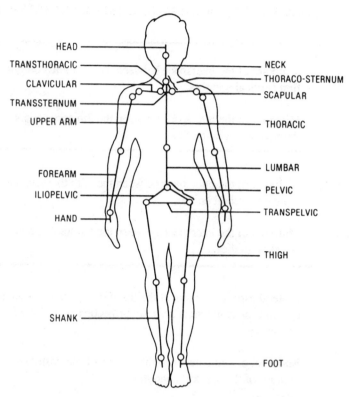

Figure 5-4. Typical link-joint system (NASA/Webb 1978).

is much more difficult for complex joints with several degrees of freedom, such as in the hip or, even more difficult, in the shoulder. In fact, the shoulder joint is depicted in Figure 5-4 as two joints with an intermediate scapular link. In many cases, the biomechanicist will simply but carefully assume location and properties of a joint in question so that it fits the given model requirements; hence, the model joint characteristics may reflect the true body articulation only partially.

Once the joints are established, a straight-line link length is defined as the distance between adjacent joint centers. Tables 5-1 and 5-2 list the definitions of the joint centers and of the links between them.

Unfortunately, anthropometric dimensions are not measured between joints but usually between externally discernible landmarks, such as bony protrusions on the skeletal system. Hence, a major problem in developing a model depicting the human body is to establish the numerical relationships between standard anthropometric measures and link lengths. This usually cannot be done by simply expressing segment lengths as proportions of body stature, a procedure occasionally and then usually falsely propagated. As the correlation tables in Chapter 1 reflect, stature is highly related to some other height measures but shows only very low correlation coefficients with most other measures of the human body. Hence, in many cases one must derive link lengths from measures of bone lengths, which unfortunately are not regularly taken in anthropometric surveys. However, if such data are available, many link lengths can be obtained from regression equations — see Table 5-3. (Other means to estimate length links are discussed by Chaffin and Andersson 1984 and by NASA/Webb 1978.) Obviously, the determination

Table 5-1. Definitions of the joint centers (adapted from NASA/Webb 1978).

HEAD	
Head/Neck	Midpoint of the interspace between the occipital condyle and the first cervical vertebra.
Neck/Thorax	Midpoint of the interspace between the 7th cervical and 1st thoracic vertebral bodies.
TRUNK	
Thorax/Lumbar	Midpoint of the interspace between the 12th thoracic and 1st lumbar vertebral bodies.
Lumbar/Sacral	Midpoint of the interspace between the 5th lumbar and 1st sacral vertebral bodies.
LEG	
Hip	(Lateral aspect) A point at the tip of the femoral trochanter 1.0 cm anterior to the most laterally projecting part of the femoral trochanter.
Knee	Midpoint of a line between the centers of the posterior convexities of the femoral condyles.
Ankle	Level of a line between the tip of the lateral malleolus of the fibula and a point 5 mm distal to the tibial malleolus.
ARM	
Sternoclavicular	Midpoint position of the palpable junction between the proximal end of the clavicle and the sternum at the upper border (jugular notch) of the sternum.
Claviscapular	Midpoint of a line between the coracoid tuberosity of the clavicle (at the posterior border of the bone) and the acromioclavicular articulation (or the tubercle at the lateral end of the clavicle); the point should be visualized as on the underside of the clavicle.
Glenohumeral	Midregion of the palpable bony mass of the head and tuberosities of the humerus; with the arm abducted about 45° relative to the vertebral margin of the scapula, a line dropped perpendicular to the long axis of the arm from the outermost margin of the acromion approximately bisects the joint.

Table 5-2. Definition of links (adapted from NASA/Webb 1978).

HEAD

The straight line between the occipital condyle/C1 interspace center and the center of mass of the head.

NECK

The straight line between the occipital condyle/C1 and the C7/T1 vertebral interspace joint centers.

TORSO (total)

The straight-line distance from the occipital condyle/C1 interspace joint center to the midpoint of a line passing through the right and left hip joint centers.

THORAX Sub-Link

Thoraco-sternum —— A closed linkage system composed of three links. The right and left transthorax are straight-line distances from the C7/T1 interspace to the right and left sternoclavicular joint centers. The transternum link is a straight-line distance between the right and left sternoclavicular joint centers.

Clavicular —— The straight-line between the sternoclavicular and the claviscapular joint centers.

Scapular —— The straight-line between the claviscapular and glenohumeral joint centers.

Thoracic —— The straight-line between C7/T1 and T12/L1 vertebral body interspace joint centers.

LUMBAR Sub-Link

The straight-line between the T12/L1 and L5/S1 vertebrae interspace joint centers.

PELVIS Sub-Link

Treated as a system composed of three links: the right and left iliopelvic links are straight-lines between the L5/S1 interspace joint center and a hip joint center. The transpelvic link is a straight-line between the right and left hip joint centers.

THIGH

The straight-line between the hip and knee joint centers of rotation.

SHANK

The straight-line between the knee and akle joint centers of rotation.

FOOT

The straight-line between the ankle joint center and the center of mass of the foot.

UPPER ARM

The straight-line between the glenohumeral and elbow joint centers of rotation.

FOREARM

The straight-line between the elbow and wrist joint centers of rotation.

HAND

The straight-line between the wrist joint center of rotation and the center of mass of the hand.

Table 5-3. Regression equations for estimating link lengths (in cm) directly from bone lengths (adapted from NASA/Webb 1978).

Empirical Equation		Standard Error of Estimate	Correlation Coefficient
Forearm Link Length	= 1.0709 Radius Length	NA	NA
	= 0.9870 Ulna Length	NA	NA
Arm Link Length	= 66.2621 + 0.8665 Ulna Length	9.90	0.94
	= 58.0752 + 0.9646 Radius Length	8.92	0.94
Thigh Link Length	= 132.8253 + 0.8172 Tibia Length	16.57	0.73
	= 92.0397 + 0.8699 Fibula Length	10.34	0.87
Shank Link Length	= 8.2184 + 1.0904 Fibula Length	5.95	0.97
Shank Link Length	= 1.0776 Tibia Length	NA	NA

of link lengths from anthropometric data is still a difficult task for which both methodologies and data need to be developed.

Assessment of Body Volumes

Measurement of whole body or of segment volume is often necessary to calculate inertial properties or to design close-fitting garments. Use of the "Archimedes Principle" provides the body volume; the subject is immersed in a container filled with water, and the displaced water yields the volume. The technique works well for obtaining volumes of limbs and also of the whole body if changes due to respiration can be controlled.

There are also indirect methods to obtain the volume. One is to use information about the cross-section contours, which can be obtained by CAT scans or by dissection. If these cross-section contours are taken at sufficiently close separations so that the changes between cross-sections can be assumed linear with distance, the volume V can be calculated by summing the cross-section areas A multiplied with their distances d from each other:

$$V = \Sigma(A_i \, d_i) \tag{5-10}$$

Often, the distance d between cross-section costs is kept constant, and adjacent cross-sectional areas are averaged:

$$\overline{A}_i = \frac{1}{2}\Sigma(A_{i-1} + A_i) \tag{5-11}$$

Other approximations rely on the assumption that body segments resemble regular geometric figures. For example, if the body cross-section is elliptical, then it can be described by

$$A_i = \pi \, a_i b_i \tag{5-12}$$

where a is the semimajor axis and b is the semiminor axis. The volume can then be calculated according to equation 5-10. Of course, one may assume that body segments resemble cylindrical cone sections or that they can be represented by cylinders. Such simplifying assumptions lend themselves to easy calculations of the volume.

Assessment of Inertial Properties

Knowledge of the total body mass and its distribution throughout the body is important for the assessment of dynamic properties of the human body. To obtain such data, many methods and techniques have been developed: see, e.g., Chaffin and Andersson 1984; Hay 1973; NASA/Webb 1978; Kaleps, Clauser, Young, Chandler, Zehner, and McConville 1984; McConville, Churchill, Kaleps, Clauser, and Cuzzi 1980; Roebuck, Kroemer, and Thomson 1975 for detailed compilations and discussions.

The simplest inertial property is weight, which can be measured easily with a variety of scales. (Human body weight is usually measured in air, hence there is a slight error due to buoyancy.) Using cadaver data, body segment weight can be predicted from total body weight — see Table 5-4.

According to Newton's Second Law, weight W is a force depending on body mass M and the gravitational acceleration g, according to

$$W = M g \qquad (5\text{-}13)$$

Density D is the mass per unit volume:

$$D = M \, V^{-1} = W \, g^{-1} \, V^{-1} \qquad (5\text{-}14)$$

Mass is, of course,

$$M = D V \qquad (5\text{-}15)$$

The specific density D_s is the ratio of D to the density of water, D_w

$$D_s = D \, D_w^{-1} \qquad (5\text{-}16)$$

Table 5-4. Prediction equations to estimate segment mass (in kg) from total body weight W (adapted from NASA/Webb 1978).

Segment	Empirical Equation			Standard Error of Estimate	Correlation Coefficient
Head	0.0306	W	+ 2.46	± 0.43	0.626
Head and neck	0.0534	W	+ 2.33	± 0.60	0.726
Neck	0.0146	W	+ 0.60	± 0.21	0.666
Head, neck and torso	0.5940	W	− 2.20	± 2.01	0.949
Neck and torso	0.5582	W	− 4.26	± 1.72	0.958
Total arm	0.0505	W	+ 0.01	± 0.35	0.829
Upper arm	0.0274	W	− 0.01	± 0.19	0.826
Forearm and hand	0.0233	W	− 0.01	± 0.20	0.762
Forearm	0.0189	W	− 0.16	± 0.15	0.783
Hand	0.0055	W	+ 0.07	± 0.07	0.605
Total leg	0.1582	W	+ 0.05	± 1.02	0.847
Thigh	0.1159	W	− 1.02	± 0.71	0.859
Shank and foot	0.0452	W	+ 0.82	± 0.41	0.750
Shank	0.0375	W	+ 0.38	± 0.33	0.763
Foot	0.0069	W	+ 0.47	± 0.11	0.552

The human body is not homogeneous throughout; its density varies depending on cavities, water content, fat tissue, bone components, etc. Still, in many cases it is sufficient to assume that either the body segment in question or even the whole body be of constant (average) density.

Many density data have been obtained from weight and volume of dissected cadavers. However, there are obvious difficulties associated with the use of cadaver material, such as loss of fluids, chemical changes in tissue, etc. Hence, other studies used immersion and weighing techniques to determine density values on living subjects. Recently, stereophotometry has been combined with anthropometric techniques to establish mass properties of living persons (Kaleps, Clauser, Young et al. 1984).

Another useful concept to distinguish between the compositions of different bodies is that of the Lean Body Mass or Lean Body Weight. This relies on the assumption that basic structural body components such as skin, musle, bone, etc., are relatively constant in percentage composition from individual to individual. However, the component fat varies in percentage of total mass or weight throughout the body and for different persons. This allows to express body weight W as

$$W = \text{lean body weight} + \text{fat weight} \qquad (5\text{-}17)$$

There are several techniques to determine body fat. Many use skinfold measures, where in selected areas of the body the fold thickness of loose skin is measured with a special caliper. Unfortunately, measurement of skinfold thickness is a rather difficult and not very reliable procedure since skinfolds may be grasped and compressed in different manners, they may slip from the instrument, and the pressure applied by the instrument over its measuring surfaces may be varying.

Locating the Center of Mass

The body mass can be considered as concentrated in one point in the body where its physical characteristics respond in the same way as if distributed throughout the body. The measurement of the location of the mass center is somewhat difficult with living persons because respiration causes shifts in the mass distribution, so do muscular contractions and food and fluid ingestion or excretion. Of course, there are major shifts in the location in the center of mass with changed body positions and in particular with body movements.

For the body at rest, various methods exist to determine the location of the center of mass, CM. Most rely on the principle of finding the one location where a single support would keep the body balanced. One of the simplest techniques is to place the body on a platform which is supported by two scales at precisely known support points. The body weight is then counteracted by the two forces at the support points, and with their distances from the line of action of the body weight the balancing moments can be calculated. From this, their lever arms (distances) can be determined. Figure 5-5 shows this procedure and the calculations proceed as follows:

$$a\,W_1 - b\,W - c\,W_b = 0 \quad \text{Moments about } W_2 \qquad (5\text{-}18)$$

and

$$b = d + e \qquad (5\text{-}19)$$

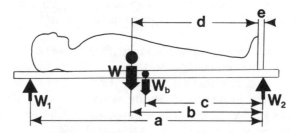

Figure 5-5. Finding the center of mass of the body placed on a known board and on two scales.

with W_1 the force at scale 1; a the distance between scales 1 and 2; W the weight of the body at its center of mass CM; W_b is the weight of the board at its CM, and c its distance from W_2; d is the distance between the CM of the body and the soles of the feet, and e their distance from W_2.

Rearranging and inserting known force and distance values provide the solution:

$$d = (aW_1 - cW_b)(W_1 + W_2 + W_b)^{-1} - e \qquad (5\text{-}20)$$

Note that the actual weight W of the subject need not be known.

This method can be used, of course, with the subject in any body position; hence, the distances in X and Z directions can be determined for the center of mass similarly to the determination in the Y direction above.

Table 5-5 lists relative locations of mass centers; Table 5-6 reflects ratios between segment and total body masses, found in several studies, some of which are more than a century old. For further information about mass properties of the human body and their assessment, see Chandler, Clauser, McConville, Reynolds, and Young 1975, Chaffin and Andersson 1984, and Roebuck, Kroemer, and Thomson 1975.

Kinematic Chain Models

The "stick person" concept consisting of links and joints, embellished with volumes and masses and driven by muscles can be used to model human motion and strength capabilities.

Figure 5-6 shows a model of the human body related to the one shown in Figure 5-4. Forces exerted with the hand to an outside object (H_x, H_y, or H_z) or the torques generated in the hand (T_H) are transmitted along the links. First, the force exerted with the right hand, modified by the existing mechanical advantages, must be transmitted across the right elbow (E). (Also, at the elbow additional force or torque must be generated to support the mass of the forearm. However, for the moment, the model will be considered massless.) Similarly, the shoulder S must transmit the same efforts, again modified by existing mechanical conditions. In this manner, all subsequent joints transmit the effort exerted with the hands throughout the trunk, hips, and legs and finally from the foot to the floor. Here, the orthogonal force and torque vectors can be separated again, similarly to the vector analysis at the hands. Still assuming a massless body, the same sum of vectors must exist at the feet as was found at the hands.

Table 5-5. Locations of the centers of mass of body segments, measured in percent from their proximal ends (adapted from Roebuck, Kroemer, and Thomson 1975).

	Harless (1860)	Braune and Fischer (1889)	Fischer (1906)	Dempster (1955)	Clauser, McConville and Young (1969)
Sample Size	2	3	1	8	13
Head	36.2%	--	--	43.3%	46.6%
Trunk*	44.8	--	--	--	38.0
Total arm	--	--	44.6%	--	41.3
Upper arm	--	47.0%	45.0	43.6	51.3
Forearm and hand*	--	47.2	46.2	67.7	62.6
Forearm*	42.0	42.1	--	43.0	39.0
Hand*	39.7	--	--	49.4	18.0
Total leg*	--	--	41.2	43.3	38.2
Thigh*	48.9	44.0	43.6	43.3	37.2
Calf and foot	--	52.4	53.7	43.7	47.5
Calf	43.3	42.0	43.3	43.3	37.1
Foot	44.4	44.4	--	42.9	44.9
Total body	41.4	--	--	--	41.2

* The values on these lines are not directly comparable since the different investigators used differing definitions for segment lengths.

Table 5-6. Segmental mass ratios in percent derived from cadaver studies (adapted from Roebuck, Kroemer, and Thomson 1975).

	Harless (1860)	Braune and Fischer (1889)	Fischer (1906)	Dempster* (1955)	Clauser, McConville and Young (1969)	Average
Sample Size	2	3	1	8	13	5
Head	7.6%	7.0%	8.8%	8.1%	7.3	7.8
Trunk	44.2	46.1	45.2	49.7	50.7	47.2
Total arm	5.7	6.2	5.4	5.0	4.9	5.4
Upper arm	3.2	3.3	2.8	2.8	2.6	2.9
Forearm and hand	2.6	2.9	2.6	2.2	2.3	2.5
Forearm	1.7	2.1	--	1.6	1.6	1.8
Hand	0.9	0.8	--	0.6	0.7	0.8
Total leg	18.4	17.2	17.6	16.1	16.1	17.1
Thigh	11.9	10.7	11.0	9.9	10.3	10.8
Calf and foot	6.6	6.5	6.6	6.1	5.8	6.3
Calf	4.6	4.8	4.5	4.6	4.3	4.6
Foot	2.0	1.7	2.1	1.4	1.5	1.7
Total body**	100.0	100.0	100.0	100.0	100.0	100.0

*Dempster's values adjusted by Clauser, McConville, and Young (1969).
**Calculated from head + trunk + 2 (total arm + total leg).

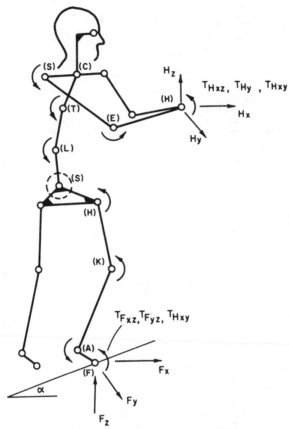

Figure 5-6. Model of the chain of forces (F) or torques (T) transmitted from the hand through arm, trunk and legs to the floor.

Of course, the assumption of no body mass is unrealistic. This can be remedied by incorporating information about mass properties of the human body as mentioned before. Furthermore, the consideration of only the efforts visible at the interfaces between the body and the environment disregards the fact that, at all body joints, antagonistic muscle groups exist which may counteract each other for control and stabilization. Their possibly high individual efforts may nullify each other for the outside observer.

Also, this example does not consider three-dimensional conditions. Finally, body motions, instead of the static position assumed here, would complicate the model.

However, if this simple model is accepted, a set of equations allows its computational analysis. These equations follow from the standard procedure of setting the sum of forces in all directions to zero and likewise the sums of all torques.

$$H_x + F_x = 0 \qquad\qquad (5\text{-}21)$$
$$H_y + F_y = 0 \qquad\qquad (5\text{-}22)$$
$$H_z + F_z = 0 \qquad\qquad (5\text{-}23)$$
$$T_{H_{xz}} + T_{F_{xz}} = 0 \qquad\qquad (5\text{-}24)$$
$$T_{H_{yz}} + T_{F_{yz}} = 0 \qquad\qquad (5\text{-}25)$$
$$T_{H_{xy}} + T_{F_{xy}} = 0 \qquad\qquad (5\text{-}26)$$

In this example, it is assumed that one foot standing on a plane inclined by the angle α counteracts all the hand efforts.

Following such procedures, rather complex models of the human body can be developed as shown in detail by Chaffin and Andersson (1984).

Description of Human Motion

Classically, human movement has been described with anatomically derived terms used in the medical profession. Unfortunately, there are systematic and deep-rooted problems associated with these terms: motions are considered to begin from the so-called anatomical position as depicted earlier in Figure 1-1, Chapter 1 on anthropometry but with the palms of the hands twisted forwards, i.e., "supinated." Words to describe motions were generated using the nouns flexion, tension, duction, rotation, and nation together with the prefixes ex, hyper, ad, ab, in, out, supi, and pro (see Figure 5-7). Unfortunately, the same

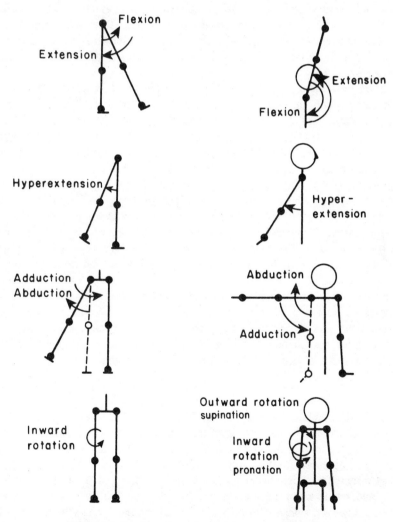

Figure 5-7. Classic notation of body motions.

terms are applied indiscriminately to rotational movements about joints and to trans-lational displacements of limbs; certain motions said to occur in a given plane may also occur in others. For example, rotation of the arm can occur in the sagittal, frontal, or transverse planes and practically in any other orientation. Severe difficulties are associ-ated with motion in the shoulder joint. For example, having "abducted" the upper arm90° sideways, the limb actually moves toward the body center line ("adduction") in whatever motion is performed. Similar problems arise when one tries to describe inward and outward rotation of the forearm. It is very difficult to describe relative locations of body parts, and a "zero position" is often not specified — see Figure 2-3 and Table 2-1 in Chapter 2.

In 1968, Roebuck made a thorough and radically new attempt to generate a more appropriate terminology (described in detail by Roebuck, Kroemer and Thomson 1975). It uses a coordinate system that is attached to the human pelvic area as shown in Figure 1-1. The common frontal, sagittal, and transverse planes are used, and the usual 360° rotation is centered at each body articulation considered. The important basic concept is to describe consistently movements of the limbs with respect to their joints, as referred to the overall coordinate reference point in the pelvic area. Roebuck's system uses the notation of a clockwise/counterclockwise (as seen by the subject) rotation and of "zero" being footward in the frontal and sagittal planes, and forward in the transverse plane — see Figures 5-8 and 5-9.

To describe movement verbally, Roebuck used the word "vection" (from the Latin *vehere*, to carry) which, unfortunately, did not find widespread acceptance in spite of the fact that it replaced both "duction" and "nation." Hence, it was proposed that the common "twist" be used to indicate rotation about a long axis ("nation"), while the familiar "flexion" and "extention" be maintained with "pivot" indicating rotation about an axis perpendicular to the flexion-extension axis (Kroemer, Marras, McGlothlin, McIn-tyre, and Nordin 1989). Thus, for example, the index finger may be flexed or extended, or pivoted ulnarly or radially. Roebuck's prefixes are kept, "e-" signifying out or up from the standard position; "in-" the opposite of e-, i.e., in or down; "posi-" and "negi-" are self-explanatory. Table 5-7 lists the terms. Any motion can be described in this terminology by a descriptive term, composed of the designations for the plane, the direction, and the type of motion. For example, "sagflexion" indicates that the motion occurs in the sagittal plane.

Consistent use of reference points, planes, and motion terms should reduce uncertainty that exists about human motion capabilities and allow its measurement and reporting utilizing computer models and systems. For data regarding current information about motion capabilities, see Chapter 2. For instruments suitable for measuring joint mobility, check Chaffin and Andersson (1984) or Roebuck, Kroemer, and Thomson (1975).

SUMMARY

Biomechanical modeling of the human body, or of its parts, normally requires simplification of actual (physiologic, anatomic, anthropometric, etc.) characteristics to fit the methods and techniques derived from mechanics. While this establishes limitations regarding the completeness, reliability, and validity of the biomechanical procedures, it also allows research and conclusions with unique insights which would not have been possible to understand when using traditional (physiologic, anatomic, or anthropometric) approaches. However, one has to be keenly aware of the limitations imposed by the underlying simplifications.

- An important application of biomechanics is in the calculation of torques or forces that can be developed by muscles about body joints. In reversing this procedure, one can assess the strain on muscles, bones, and tissues generated by external loads on the body in various positions.
- Body segment dimensions, their volumes and mass properties can be calculated from anthropometric data.
- Kinematic chain models of linked body segments allow the prediction of total-body capability (e.g., lifting) from the consideration of body segment capabilities.
- A taxonomy for position and motion of the body exists which describes these both mathematically and verbally.

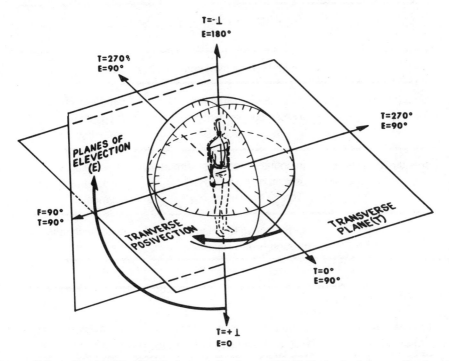

Figure 5-8. Roebuck's "global coordinate system" (adapted from Roebuck, Kroemer, and Thomson 1975).

Table 5-7. New Terminology to Describe Body Motions.

New Terms	Meaning	Replacing
pivot and flexion, extension	rotation of a body segment abouts its proximal joint	duction vection rotation
twist	rotation of a body segment about its internal axis	-nation rotation
ex-	away (up, out) from zero	ab- e-
in-	towards (in, down) zero	ad-
clock(wise) (or none)	in clockwise direction, as seen on own body	supi- or pro- (depending on body segment)
counter-(clockwise)	in counter-clockwise direction, as seen on own body	pro- or supi- (depending on body segment
Front-, Trans-, Sag-	Reference to the plane in which motion is described	uncertainty and confusion

Examples:

Front-ex-pivot	=	pivoting movement in the frontal plane, away from zero
Trans-in-twist	=	twisting movement in the transverse plane, clockwise

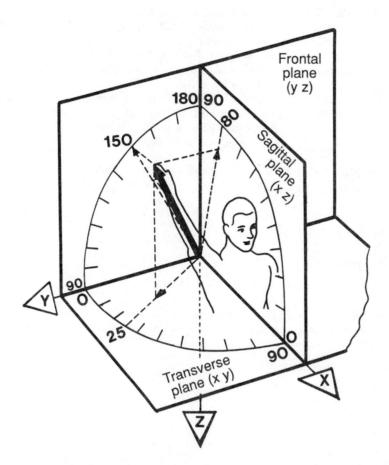

Figure 5-9.   Description of the angular position of the right arm in the new notation system (adapted from Roebuck, Kroemer, and Thomson 1975).   The position of the hand point on the arm vector is F = 150°, S = 170°, T = 65°.   Note that the directions for x, y, and z are commonly used in biomechanics (NASA/ Webb, 1978) to describe accelerations of the "vehicle" that carries a human occupant.

REFERENCES

Chaffin, D. B. and Andersson, G. B. J. 1984. *Occupational Biomechanics*. New York, NY: Wiley.

Chandler, R. F., Clauser, C. E., McConville, J. R., Reynolds, H. M., and Young, J. W. 1975. Investigation of Inertial Properties of the Human Body. AMRL-TR-74-137, Wright-Patterson AFB, OH: Aerospace Medical Research Laboratory.

Hay, J. G. 1973. The Center of Gravity of the Human Body. In *Kinesiology III* (pp. 20–44). Washington, DC: American Association for Health, Physical Education, and Recreation.

Kaleps, I., Clauser, C. E., Young, J. W., Chandler, R. F., Zehner, G. F., and McConville, J. 1984. Investigation Into the Mass Distribution Properties of the Human Body and its Segments. *Ergonomics,* 27:12:1225–1237.

King, A. I. 1984. A Review of Biomechanical Models. *Journal of Biomechanical Engineering,* 106:97–104.

Kroemer, K. H. E., Marras, W. S., McGlothlin, J. D., McIntyre, D. R., and Nordin, M. 1989. Assessing Human Dynamic Muscle Strength. (Technical Report, 8-30-89). Blacksburg, VA: Virginia Tech, Industrial Ergonomics Laboratory.

Kroemer, K. H. E., Snook, S. H., Meadows, S. K., and Deutsch, S., (eds), 1987. *Ergonomic Models of Anthropometry, Human Biomechanics and Operator-Equipment Interfaces*. Washington, D.C.: National Academy of Sciences.

McConville, J. T., Churchill, T., Kaleps, I., Clauser, C. E., and Cuzzi, J. 1980. Anthropometric Relationships of Body and Body Segment Moments of Inertia. AFAMRL-TR-80-119, Wright-Patterson AFB, OH: Aerospace Medical Research Laboratory.

NASA (1987). Man System Interpretation Standards. (NASA-ST-3000). Houston, TX: L. B. J. Space Center.

NASA/Webb (Eds.) 1978. *Anthropometric Sourcebook* (3 volumes). NASA Reference Publication 1024. Houston, TX: L.B.J. Space Center, NASA (NTIS, Springfield, VA 22161, Order No. 79 11 734).

Roebuck, J. A., Kroemer, K. H. E., and Thomson, W. G. 1975. *Engineering Anthropometry Methods*. New York, NY: Wiley.

FURTHER READING

Chaffin, D. B. and Andersson, G. B. J. 1984. *Occupational Biomechanics*. New York, NY: Wiley.

Kroemer, K. H. E., Snook, S. H., Meadows, S. K., and Deutsch, S. (eds). 1987. Ergonomic Models of Anthropometry, Human Biomechanics and Operator-Equipment Interfaces. Washington, D.C.: National Academy of Sciences.

Journals: e.g.
  Applied Ergonomics
  Biomechanical Engineering
  Biomechanics
  Ergonomics
  Human Factors
  Kinesiology
  Spine

Chapter 6

THE RESPIRATORY SYSTEM

OVERVIEW

The respiratory system provides oxygen for the energy metabolism and dissipates metabolic byproducts. In the lungs, oxygen is absorbed into the blood. The circulatory system transports oxygen (and nutrients) throughout the body, particularly to the working muscles. It removes metabolic byproducts either to the skin (where water and heat are dissipated) or to the lungs. There, carbon dioxide as well as water and heat are dispelled while more oxygen is absorbed.

The Model

Figure 6-1 indicates the close interaction between the respiratory system, which absorbs oxygen and dispells carbon dioxide, water and heat, and the circulatory system which provides the transport means.

INTRODUCTION

The respiratory system moves air to and from the lungs, where part of the oxygen contained in the inhaled air is absorbed into the blood stream; it also removes carbon dioxide, water, and heat into the air to be exhaled. The absorption of oxygen and the expulsion of metabolites takes place at spongelike surfaces in the lungs that contain small air sacs (*alveoli*).

ARCHITECTURE

The so-called "respiratory tree" in Figure 6-2 shows schematically the path of the air that has entered the body via the nose and mouth and passed through the *pharynx* (throat), *larynx* (voice box), and the *trachea* (windpipe) into the bronchial tree. Here, the airways branch repeatedly until they terminate in the alveoli of the lungs. These are small air sacs with diameters of about 400 μm, with thin membranes between capillaries and other subcellular structures allowing the exchange of O_2 and CO_2 between blood and air.

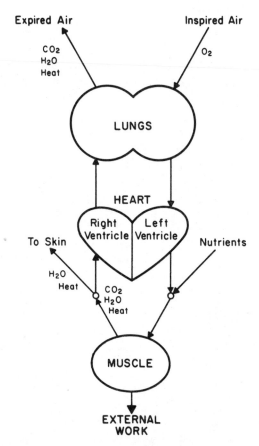

Figure 6-1. Scheme of the interrelated functions of the respiratory and circulatory systems.

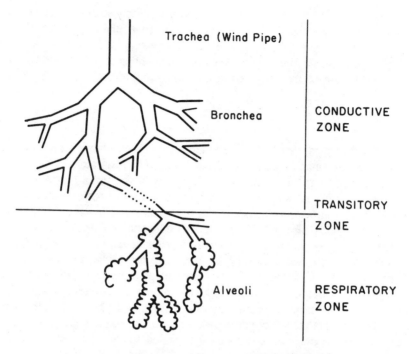

Trachea (Wind Pipe)

Bronchea

CONDUCTIVE
ZONE

TRANSITORY
ZONE

Alveoli

RESPIRATORY
ZONE

Figure 6-2. Scheme of the "respiratory tree".

The adult has twenty-three subsequent dichotomic branching steps. The first sixteen branches merely conduct the air with little or no gas exchange. The next three branchings are respiratory bronchioles (diameter of about 1 mm) with some gas exchange. The final four generations of branches form the alveolar ducts which terminate in the alveoli. Here, most of the gas exchange takes place: the alveolar ducts and sacs are fully alveolated, while the respiratory bronchioles have variously spaced alveoli. Altogether, between 200 and 600 million alveoli provide a grown person with about 70 to 90 m^2 of exchange surface.

The air exchange is brought about by the pumping action of the thorax. The diaphragm separating the chest cavity from the abdomen descends about 10 cm when the abdominal muscles relax (which brings about a protrusion of the abdomen). Furthermore, muscles connecting the ribs contract and, by their anatomical and mechanical arrangements, raise the ribs. Hence, the dimensions of the rib cage and of its included thoracic cavity increase both towards the outside and in the direction of the abdomen: air is sucked into the lungs. When the inspiratory muscles relax, elastic recoil in lung tissue, thoracic wall, and abdomen restores their resting positions without involvement of expiratory muscles: air is expelled from the lungs. However, when ventilation needs are several times higher than the resting value, such as with heavy work, the recoil forces are augmented by activities of the expiratory muscles. These internal intercostal muscle fibers have a direction opposite those of the external intercostal muscles and reduce the thoracic cavity. Furthermore, contraction of the muscles in the abdominal wall can also assist expiration.

Thus, inspiratory and expiratory muscles are activated reciprocally, and both overcome the resistance provided by the elastic properties of the chest wall and of pulmonary tissue, particularly of the airways: of the total resistance, about 4/5 is airway resistance and the remainder tissue resistance. The high airway resistance is due to the fact that the air flow in the trachea and the main bronchi is turbulent, particularly so at the high flow velocities required by heavy exercise. However, in the finest air tubes, of which there are very many and together have a large volume, air flow is slow and laminar. Altogether, the energy required for breathing is relatively small, amounting to only about 2 percent of the total oxygen uptake of the body at rest and increasing to not more than 10 percent at heavy exercise.

FUNCTIONS

Of course, the primary task of the respiratory system is to move air through the lungs, which provide exchange of gases, heat, and water. The respiratory system also "conditions" the inspired air: it adjusts the temperature of the inward flowing air to body temperature, moistens or dries the air, and cleanses it from particles. All this takes place at mucus-covered surfaces in the nose, mouth, and throat. The temperature regulation is so efficient that the inspired air is at about body core temperature, $37^{\circ}C$, when it reaches the end of the pharynx, whether inspiration is through the mouth or through the nose.

In a normal climate, about 10% of the total heat loss of the body, whether at rest or work, occurs in the respiratory tract. This percentage increases to about 25% at outside temperatures of about $-30^{\circ}C$. In a cold environment, heating and humidifying the inspired air cools the mucosa; during expiration, some of the heat and water is recovered by condensation from the air to be exhaled. (Hence the "runny nose" in the cold.) Thus, the respiratory tract not only conditions the inhaled air but also recovers some of the spent energy when the air is exhaled.

Also, the respiratory tract cleans the air from particles of foreign matter. Particles larger than about 10 μm are trapped in the moist membranes of the nose. Smaller particles settle in the walls of the trachea, the bronchi, and the bronchioles. Thus, the lungs are kept practically sterile, and an occasional sneeze or cough (with air movements at approximately the speed of sound in the deeper parts of the respiratory system) helps to expel foreign particles.

Respiratory Volumes

The volume of air exchanged in the lungs depends much on muscular activities, that is on the requirements associated with the kind of work performed. When the respiratory muscles are relaxed, there is still air left in the lungs. A forced maximal expiration reduces this amount of air in the lungs to the so-called "residual volume" (or residual "capacity"), see Figure 6-3. The following maximal inspiration adds the volume called "vital capacity." Both volumes are the "total lung capacity." During rest or submaximal work, only the so-called "tidal volume" is moved, leaving both an inspiratory and an expiratory "reserve volume" within the vital capacity. ("Dead volume," or "dead space" is the volume — of about 0.2 L — of the conductive zone of the human airways where the air does not come in contact with alveoli and, therefore, does not contribute to the gas exchange.)

Vital capacity and other respiratory volumes are usually measured with the help of a spirometer. The results depend on the age, training, sex, body size, and body position of the subject. Total lung volume for highly trained tall young males is between 7 and 8 L and their vital capacity up to 6 L. Women have lung volumes that are about 10% smaller than their male peers. Untrained persons have volumes of about 60% to 80% of their athletic peers.

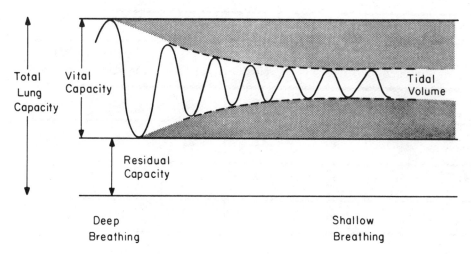

Figure 6-3. Respiratory volumes.

"Pulmonary ventilation" is the movement of gas in and out of the lungs. The pulmonary ventilation is calculated by multiplying the frequency of breathing by the expired tidal volume. This is called the (respiratory expired) "minute volume." At rest, one breathes 10 to 20 times/min. In light exercise, primarily the tidal volume is increased, but with heavier work the respiratory frequency also quickly increases up to about 45 breaths/min. (Children breathe much faster at maximal effort — about 55 times/min at 12 years of age and at about 70 times/min at 5 years.) This indicates that breathing frequency, which can be measured easily, is not a very reliable indicator of the heaviness of work performed.

Table 6-1 presents an overview of gas partial pressures at rest, under certain simplifying assumptions such as completely dry inspired air. The efficiency of gas exchange in the lungs is indicated by the decreased partial oxygen pressure and increased partial pressure of CO_2 in the expired air as compared to the inhaled air. The gas pressures in the venous blood depend very much on the metabolic activities of the tissues. For example, relatively more oxygen is used in the muscle and more carbon dioxide produced then in other tissues such as kidneys and skin. This becomes apparent as one compares the gas tensions in venous blood that is either mixed after having passed through various tissues or that is observed after passage through a muscle. Of course, the gas tensions shown in this table for resting individuals change much during strenuous work. For example, the carbon dioxide tension in mixed venous blood at maximal exercise may be in the neighborhood of 70 mmHg as compared to 47 at rest; accordingly, the partial pressure of oxygen may be reduced to 17 from 40 mmHg.

As compared to rest, the respiratory system is able to increase its moved volumes and absorbed oxygen by large multiples. The minute volume can be increased from about 5 L/min to 100 L/min or more; that is an increase in air volume by a factor of 20 or more. Though not exactly linearly related to it, the oxygen consumption shows a similar increase.

Table 6-1. Gas partial pressures at rest, at 760 mmHg, with dry inspired air. (Based on Astrand and Rodahl 1977, Rodgers 1986).

	AIR			BLOOD		
	Inspired	Alveolar	Expired	Arterial	Muscle Capillary	Mixed Venous
O_2	150	100	116	95	30	40
CO_2	0.3	40	32	40	50	47
N_2	610	573	565	578	573	576
H_2O	0	47	47	47	47	47

SUMMARY

- Air entering the respiratory tree is conditioned so that gas exchange at the alveoli is facilitated: O_2 is absorbed into the blood, and CO_2, heat, and water are dispelled into the air to be exhaled.
- Only a part (the tidal volume) of the available volume (vital capacity) of the airways is normally utilized for respiratory exchange.
- The respiratory system can easily increase its moved air volume by a factor of 20 over resting conditions which is accompanied by a similar increase in oxygen intake, needed to perform physically demanding work.

REFERENCES

Astrand, P. 0. and Rodahl, K. 1977. *Textbook of Work Physiology*. Second edition. New York, NY: McGraw-Hill.
Rodgers, S. R. 1986. Personal Communication, 24 March 1986.

FURTHER READING

Astrand, P. 0. and Rodahl, K. 1986. *Textbook of Work Physiology*. Third edition. New York, NY: McGraw-Hill.
Guyton, A. C. 1979. *Physiology of the Human Body*. Fifth edition. Philadelphia, PA: Saunders.
Stegemann, J. 1981. *Exercise Physiology*. Chicago, IL: Yearbook Medical Publications, Thieme.
Weller, H. and Wiley, R. L. 1979. *Basic Human Physiology*. New York, NY: Van Nostrand Reinhold.

Chapter 7

THE CIRCULATORY SYSTEMS

OVERVIEW

Two systems transport materials between body cells and tissues: the blood and the lymphatic system. They take nutritional materials from the digestive tract to the cells for catabolism, synthesis, and deposit. The blood provides oxygen by carrying it from the lungs to the consuming cells, and it carries metabolites: carbon dioxide is brought to the lungs to be expelled, lactic acid is transferred to the liver and kidneys for processing, and heated body fluids, mainly water, are taken to the skin and lungs for heat dissipation. Furthermore, blood is part of the body control system by carrying hormones from the endocrine glands to the cells.

The Model

The circulatory system carries oxygen from the lungs to the cells where nutritional materials, also brought by circulation from the digestive tract, are metabolized. Metabolic byproducts (CO_2, heat, and water) are dissipated by circulation. The circulatory and respiratory systems are closely interrelated as shown earlier in Figure 6-1.

BODY FLUIDS

Water is the largest weight component of the body: about 60% of body weight in men, about 50% in women. In slim individuals, the percentage of total water is higher than in obese persons since adipose tissue contains very little water. The relation between water and lean (fat-free) body mass is rather constant in "normal" adults, at about 72%. Of the total body water (say 40 L in an adult) about 70% (25 L) is contained within body cells (*intracellular* fluid); separated by the cell membrane, the other 30% (about 15 L) surround the cells (*extracellular* fluid). Extracellular fluid has an ionic composition similar to sea water. The concentrations of various electrolytes are different in the extra- and intracellular fluids. The cell membrane, which separates those two fluids, acts as a barrier for positively charged ions (cations).

Of the extracellular fluid, some is contained within blood vessels (*intravascular*) while the rest is between the blood vessels and the cell membrane (*interstitial fluid*); the cells "bathe" in interstitial fluid. Substance exchange between the intravascular and interstitial fluids takes place primarily through the walls of the capillaries; however, plasma and blood corpuscles cannot pass through these walls.

Blood

Approximately 10% of the total fluid volume consists of blood. The total volume of blood is variable, however, depending on age, gender, and training. Astrand and Rodahl (1977) consider volumes of 4 to 4.5 L of blood for women and 5 to 6 L for men as normal. Of this, about 2/3 are usually located in the venous system with the rest in the arterial vessels. The specific heat of blood is 3.85 J (0.92 cal) per gram.

Of the total blood volume, about 55% is plasma, which consists of 10% solids and 90% water. The remaining 45% of the blood volume consist of formed elements (solids). These are predominantly red cells (erythrocytes), white cells (leukocytes), and platelets (thrombocytes). As per volume, blood consists mostly of a mixture of plasma and red blood cells. The percentage of red cell volume in the total blood volume is called hematocrit.

Blood Groups

According to the content of certain antigens and antibodies, blood is classified into four groups: A, B, AB, 0. The importance of these classifications lies primarily in their incompatibility reactions in blood transfusions. However, there are other subdivisions, in particular the one according to the rhesus (Rh) factor. This is important as an obstetrical problem which occurs when a pregnant rhesus-negative woman is carrying the child of a rhesus-positive father.

Functions

The plasma carries dissolved materials to the cells, particularly oxygen and nutritive materials (monosaccharides, neutral fats, amino acids — see Chapter 8) as well as hormones, enzymes, salts, and vitamins. It removes waste products, particularly dissolved carbon dioxide and heat.

The red blood cells perform the oxygen transport. Oxygen attaches to hemoglobin, an iron-containing protein molecule of the red blood cell. Each molecule of hemoglobin contains four atoms of iron which combine loosely and reversibly with four molecules of oxygen. Carbon dioxide molecules can be bound to amino acids of this protein; since these are different binding sites, hemoglobin molecules can react simultaneously with oxygen and carbon dioxide. (*Note*: there is high affinity of hemoglobin to carbon monoxide which takes up spaces otherwise taken by oxygen; this explains the high toxicity of CO.) Each gram of hemoglobin can combine, at best, with 1.34 mL of oxygen. With 150 grams hemoglobin per liter of blood, a fully O_2-saturated liter of blood can carry 0.20 L of O_2; also, about 0.003 L O_2 is dissolved in the plasma. The approximately 5 L of blood in the body circulate through it in about one minute when the body is at rest.

THE LYMPHATIC SYSTEM

The lymphatic system is separate from the blood vessels. Lymphatic capillaries are found in most tissue spaces. They combine into larger and larger vessels that finally lead to the neck where they empty into the blood circulation at the juncture of the left internal jugular and left subclavian veins. Thus, the lymphatic system is an accessory to the blood flow system.

The lymphatic capillaries are so small and thin-walled and thus so permeable that even very large particles and protein molecules can pass directly into them. Hence, the fluid in the lymphatics is really an overflow from tissue spaces; lymph is very much the same as interstitial fluid. The flow of lymph within its tubing system is dependent on the interstitial fluid pressure: the higher the pressure, the larger the lymph flow. The second factor affecting lymph flow is the so called lymphatic pump. After a lymph vessel is stretched by excess lymph, it automatically contracts. This contraction pushes the lymph past lymphatic valves, which allow flow only in the direction towards circulation and not backwards toward the tissue. The contractions occur periodically, about once every six to ten seconds. Finally, lymph can also be pumped by the motion of tissues surrounding the lymph vessels, such as the contraction of skeletal muscle surrounding a vessel.

These lymph-pumping mechanisms in combination generate a partial vacuum in the tissues so that excess fluid can be collected and returned through the lymphatic system to the circulation. Lymph flow is highly variable, on the average 1 to 2 mL/min, a small but sufficient amount to drain excess fluid and especially excess protein that otherwise would accumulate in the tissue spaces.

THE CIRCULATORY SYSTEM OF THE BLOOD

Introduction

Working muscles and other organs requiring blood supply, removal of metabolites, and control through the hormonal system, are located throughout the body. Hence, very different and huge demands are placed on the transport system, the circulatory system of the blood.

Architecture of the Circulatory System

The circulatory system is nominally divided into two subsystems: the systemic and the pulmonary circuits, each powered by one half of the heart (which can be considered a double pump). The left side of the heart supplies the systemic circulation, which runs from the arteries through the arterioles and capillaries to the metabolizing organ (e.g., muscle); from there through venules and veins to the heart's right side. The pulmonary system starts at the right ventricle which powers the blood flow through pulmonary artery, lungs, and pulmonary vein to the left side of the heart.

Each of the halves of the heart has an antechamber (*atrium*) and the chamber (*ventricle*), the pump proper. The atria receive blood from the veins which is then brought into the ventricles through valves. The heart is in essence a hollow muscle which produces, via contraction and with the aid of valves, the desired blood flow. The total vascular system consists of a large number of parallel, serial, and often interconnected circulatory sections which supply individual organs.

The Heart as Pump

The mechanisms for excitation and contraction of the heart muscle are quite similar to those of skeletal muscle; however, specialized cardiac cells (the sinoatrial nodes) serve as "pacemakers". They determine the frequency of contractions by propagating stimuli to other cells of the heart muscle. The heart has its own intrinsic control system which operates, without external influences, at (individually different) 50 to 70 beats/min. Changes in heart action stem from the central nervous system, which influences the heart through the sympathetic and the parasympathetic subsets of the autonomic system. Stimulation towards increase of heart action comes through the sympathetic system, mostly by increasing the heart rate, the strength of cardiac contraction, and the blood flow through the coronary blood vessels supplying the heart muscle. The parasympathetic system causes a decrease of heart activities, particularly in reducing heart rate, contraction of the atrial muscle, in conduction of impulses (lengthening the delay between atrial and ventricular contraction), and by decreasing blood flow through the coronary blood vessels. The parasympathetic system is dominant during rest periods. Thus, the sympathetic and parasympathetic nervous systems are another example of the coordinated action of two opposing control systems, found so often in the body.

The myocardial action potentials can be recorded in their algebraic sum by an electrocardiogram (ECG). The different waves observed in the ECG have been given alphabetic identifiers: the P wave is associated with the electrical stimulation of the atrium while the Q, R, S, and T waves are associated with ventricular events. The ECG is mostly employed for clinical diagnoses; however, with appropriate apparatus it can be used for counting and recording the heart rate. Figure 7-1 shows the electrical, pressure, and sound events during a contraction-relaxation cycle of the heart.

The ventricle is filled through the valve-controlled opening from the atrium. The heart muscle contracts (*systole*) and when the internal pressure is equal to the pressure in the aorta, the aortic valves open and the blood is ejected from the heart into the systemic system. Continuing contraction of the heart increases the pressure further, since not the same volume of blood can escape from the aorta as the heart presses into it. Part of the excess volume is kept in the aorta and its large branches which act as a "windkessel," an elastic pressure vessel. Then, the aortic valves close with the beginning of the relaxation (*diastole*) of the heart, and the elastic properties of the aortic walls propel the stored blood into the arterial tree where the elastic blood vessels smooth out the waves of blood volume. At rest, about half the volume in the ventricle is ejected (stroke volume) while the other half remains in the heart (residual volume). Under exercise load, the heart ejects a larger portion of the contained volume. When much blood is required but cannot be supplied, such as during very strenuous physical work with small muscle groups or during maintained isometric contractions, the heart rate can become very high.

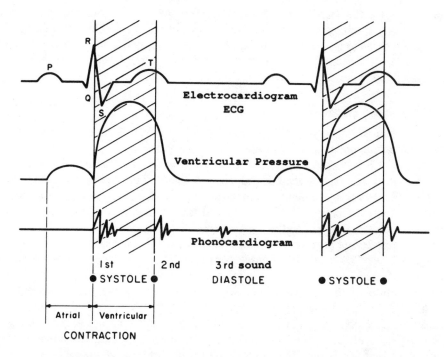

Figure 7-1. Scheme of the electrocardiogram, the pressure fluctuation, and the phonogram of the heart with its three sounds (adapted from Guyton 1979).

At a heart rate of 75 beats/min, the diastole takes less than 0.5 seconds and the systole just over 0.3 seconds; at a heart rate of 150 beats/min, the periods are close to 0.2 seconds each. Hence, an increase in heart rate occurs mainly by shortening the duration of the diastole.

The events in the right heart are similar to those in the left, but the pressures in the ventricular and pulmonary arteries are only about 1/5 of those during systole in the left heart.

Cardiac Output

The cardiac output can be affected by two factors: the frequency of contraction (heart rate) and the pressure generated by each contraction in the blood. Both determine the so-called (cardiac) *minute volume*. (The available blood volume does not vary.) The cardiac output of an adult at rest is around 5 L/min. When performing strenuous exercise, this level might be raised five times to about 25 L/min, while a well-trained athlete may reach up to 35 L/min. The ability of the heart to adjust its minute output volume to the requirements of the activity depends on two factors: on the effectiveness of the heart as a pump, and on the ease with which blood can flow through the circulatory system and return to the heart. A healthy heart can pump much more blood through the body than usually needed. Hence, an output limitation is more likely to lie in the transporting capability of the vascular portions of the circulatory system than in the heart itself. Of the total vascular system, the arterial section (before the metabolizing organ) has relatively strong elastic walls which act as a pressure vessel (windkessel) thus transmitting pressure waves far into the body, though with much loss of pressure along the way. At the arterioles of the consumer organ, the blood pressure is reduced to approximately one-third its value at the heart's aorta. Figure 7-2 shows the smoothing of pressure waves and reduction of pressure along the blood pathway schematically. As blood seeps through the consuming organ (e.g., a muscle) via capillaries, pressure drops. The pressure differential (positive on the arterial side, negative on the venous side) helps to maintain blood transport through the "capillary bed."

The "Capillary Bed"

The exchanges of oxygen, nutrients, and metabolic byproducts between the working muscle and the blood take place primarily at the "capillary bed," shown schematically in Figure 7-3. Blood enters through the *arteriole* which is circumvoluted by smooth muscles which contract or relax in response to stimuli from the symphathetic nervous system and to local accumulation of metabolites. The following *metarteriole* has fewer enclosing muscle fibers, and the entrance to the capillaries may be closed by other ring-like muscles, the pre-capillary *sphincters*. The metarteriolic and the pre-capillary sphincter muscles are predominantly controlled by local tissue conditions. Contraction or relaxation of the flow-controlling muscles change the flow resistance and hence the blood pressure. If lack of oxygen, or accumulation of metabolites, requires high blood flow, the muscles will allow the pathways to remain open; blood may even use a shortcut established by a shunt, arterial-venule anastomosis. Large cross-sectional openings in the system reduce blood flow velocity and blood pressure, allowing nutrients and oxygen to enter the extracellular space of the tissue, permitting the blood to accept metabolic byproducts from the tissue. (However, if contraction of striated muscle compresses the fine blood vessels, flow may be hindered or shut off. This is of particular consequence in sustained strong isometric contraction.) Constriction of the capillary bed reduces local blood flow so that other organs in more need of blood may be better supplied.

The venous portion of the systemic system has a large cross-section and provides low flow resistance; only about one-tenth of the total pressure loss occurs here. (This low pressure system is often called "capacitance" system in contrast to the arterial "resistance" system.) Valves are built into the venous system, allowing blood flow only toward the right ventricle. The pulmonary circulation has relatively little vessel constriction or shunting.

Hemodynamics

As in the physics of "regular" fluid dynamics, the important factors in the dynamics of the blood flow (hemodynamics) are the capacity of the pump (the heart) to do the work, the physical properties of the fluid (blood) to be pumped, particularly its viscosity ("internal friction"), and the properties of the pipes (blood vessels) with respect to the required flow rates (volume per time) and flow velocities.

The pressure gradient (the difference in internal pressure between the start and the end of the flow pathway considered) is the main determiner of flow rate and flow velocity but depends on the resistance to flow. They are related by the equation

$$\Delta p = Q\,R \qquad\qquad (7\text{-}1)$$

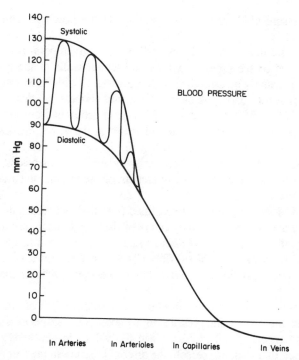

Figure 7-2. Scheme of the smoothing and reduction of blood pressure along the circulatory pathways, showing the blood pressure differential between arterial and venous sections.

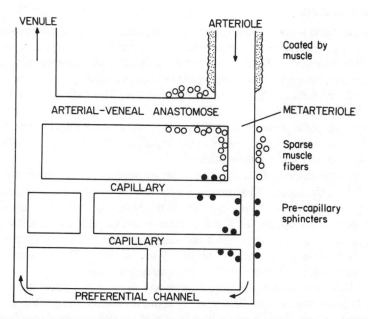

Figure 7-3. The "capillary bed" showing the branching connections between arterial and venous parts of the circulatory system, and the smooth sphincter muscles that bring about vascular constriction.

with Δp the pressure gradient (P start − P end), Q the flow rate, and R the peripheral resistance.

Since the pressure in a fluid is (theoretically) at any given point equal in all directions, it can be measured as the pressure against the lateral walls of the blood vessel. This lateral pressure is usually measured as "blood pressure," with its reading modified by the elastic properties of the containing blood vessel, and perhaps by other intervening tissues.

The flow resistance of the blood vessels is quite different from the formula used for rigid pipes but depends in essence on the diameter of the vessel, the length of the vessel considered, and on the viscosity of the blood. Blood viscosity is about 3 to 4 times greater than that of water, determined primarily by number, dimensions, and shapes of the blood cells, and dependent on the protein content of the plasma; the more hematocrit, the higher the resistance to flow.

The stream of blood may not be streamlined (laminar) but turbulent. Such turbulence may result from the fact that the outer layers of the blood stream are in physical contact with the inner walls of the blood vessels and have, therefore a much lower velocity than the more central sections of the blood flow. Turbulence increases the energy loss within the energy medium, and therefore (at a given pressure gradient) the average flow is slower.

The static pressure in a column of fluid depends on the height of that column (Pascal's Law). However, the hydrostatic pressure in, for example, the feet of a standing person is not as large as expected from physics since the valves in the veins of the extremities modify the value: in a standing person, the arterial pressure in the feet may be only about 100 mmHg higher than in the head. Nevertheless, blood, water, and other body fluids in the lower extremities are pooled there, leading to a well-known increase in volume of the lower extremities, particularly when one stands (or sits) still.

Blood Vessels

Of the blood vessels, the capillaries and to some extent also the post-capillary venules provide semi-permeable membranes to the surrounding tissue so that nutrients, gases, etc. may be exchanged. All other blood vessels, the arterioles, arteries, and veins, serve only as transport channels. The walls of these vessels consist (in different compositions) of elastic fibers, collagen fibers, and smooth muscles. The walls are thick in the big arteries and thin in the big veins. Blood flow at each point in these blood vessels is primarily determined by the pressure head of the blood wave and the diameter of the vessel at this point. Different vessels have varying capacities to influence the blood flow, blood distribution, and vascular resistance.

As already discussed, the arteries serve as pressure tanks during the ejection of the blood from the heart. Their elastic tissues stretch under the systolic impact, store this energy, and release it during diastole, thus converting an intermittent flow to a much more continuous stream. Still, the ejection of the blood from the left ventricle causes a pressure wave to travel along the blood vessels at speeds of 10 to 20 times the velocity of the blood in the aorta (which is about 0.5 ms^{-1} at rest). The frequency at which these pressure waves occur is counted as pulse rate or heart rate.

The flow resistance in the large arteries and veins is very small, the diameter of these vessels being large and the flow velocity high. In contrast, the peripheral resistance in the arterioles, metarterioles, and capillaries is high, which causes a significant reduction in blood pressure, even though the velocity of the — turbulent — blood flow is still high. The diameter of the vessels, below 0.1 mm, can be further reduced by smooth muscle fibers wrapped around the vessel, previously described. Contraction of these muscles constricts the vessels, which recoil to their normal size when the muscle relaxes. Thus, the contraction of the these smooth muscles around the blood vessels (and possibly the pressure generated by contraction of surrounding striated muscle) can change the flow characteristics and achieve reduction or complete shut-off of circulation to organs where blood supply is not so urgent, hence allowing better supply to those organs in need of it. Any increase in channel capacity (vasodilation) must be compensated by vessel constriction in other areas and/or by an increase in cardiac output, since otherwise the arterial blood pressure would fall. In the capillaries, there is no pressure variation associated with the heart's systolic and diastolic phases.

The collecting venules have a coating of connective tissue and smooth muscle rings, only intermittently spaced near the arterial bed but developing into a complete layer in their distal parts, the larger veins. Valves in the veins of extremities prevent backflow of the blood. Compression of skeletal muscle reduces the volume of veins which is restored during relaxation. Given the high flow resistance on the capillary side and the design of the valves in the veins, the blood is squeezed towards the right ventricle. Thus, the alternately contracting and relaxing muscle (doing dynamic work) acts as a pump which moves venous blood towards the right heart. The venous system usually contains about 2/3 of the total blood volume; the pulmonary veins of it contain about 15% of the total blood volume.

REGULATION OF CIRCULATION

The blood pressure in the aorta depends on cardiac output, peripheral resistance, elasticity of the main arteries, viscosity of the blood, and on the blood volume. The local flow is mainly determined by the pressure head and the diameter of the vessel through which it passes. The smooth muscles encompassing the arterioles and veins continuously receive sympathetic nerve impulses that keep the clearance (lumen) of the vessels more or less constricted. This local *vasomotor tone* is controlled by the vasoconstricting fibers driven from the medulla; the tone keeps the systemic arterial blood pressure on a level suitable for the actual requirements of all vital organs. (Changes in heart function and in circulation are initiated at brain levels above the medullary centers, probably at the cerebral cortex.) If local metabolite concentrations increase over an acceptable level, this local condition directly causes metarteriolic and sphincter muscles to relax, allowing more blood flow. In particular, if muscles require more blood flow in heavy work, signals from the motorcortex can activate vasodilation of precapillary vessels in the muscles and simultaneously trigger a vasoconstriction of the vessels supplying the abdominal organs. This leads to a remarkable and very quick redistribution of the blood supply to favor skeletal muscles over the digestive system ("muscles-over-digestion" principle).

However, even with heavy exercise the systemic blood flow is so controlled that the arterial blood pressure is sufficient for an adequate blood supply to the brain, heart, and other vital organs. To do so, neural vasoconstrictive commands can override local dilatory control. For example, the temperature-regulating center in the hypothalamus can affect vasodilation in the skin if this is needed to maintain a suitable body temperature, even if this means a reduction of blood flow to the working muscles ("skin over-muscles" principle) — see Chapter 9.

Thus, circulation at the arterial side, at the organ/consumer level is regulated both by local control and by impulses from the central nervous system, the latter having overriding power. Vasodilation (opening blood vessels beyond their clearance at regular "vasomotor tone") in the organs needing increased blood flow and vasoconstriction where blood is not so necessary, regulate local blood supply. At the same time, the heart increases its output by higher heart beat frequency; also, the blood pressure increases. At the venous side of circulation, constriction of veins, combined with the pumping action of dynamically working muscles and the forced respiratory movements, facilitate return of blood to the heart. This makes increased cardiac output possible, because the heart cannot pump more blood than it receives.

Heart rate generally follows oxygen consumption and hence energy production of the dynamically working muscle in a linear fashion from moderate to rather heavy work (see "indirect calorimetry" in Chapter 3). However, the heart rate at a given oxygen intake is higher when the work is performed with the arms than with the legs. This reflects the use of different muscles and muscle masses with different lever arms to perform the work. Smaller muscles doing the same external work as larger muscles are more strained and require more oxygen.

Also, static (isometric) muscle contraction increases the heart rate, apparently because the body tries to bring blood to the tensed muscles. However, it is difficult to compare this effect in terms of efficiency (e.g., in "beats per effort") with the increase in heart rate during dynamic efforts because, in the isometric case, there is no "work" done (in the physical sense, there is force but no displacement) while in the dynamic case, work is done.

Of course, work in a hot environment causes a higher heart rate than at a moderate temperature, as explained in Chapter 9. Finally, emotions, nervousness, apprehension, and fear can affect the heart rate at rest and during light work.

SUMMARY

- The blood flow per minute within the circulatory system depends on the actual operating characteristics of both the heart (as pump) and of the blood vessels that contain and guide the flow of the blood. The volume of blood pumped per minute is in the neighborhood of 5 L at rest, and may increase 5 or even 7 fold at strenuous work. This increase is brought about mostly by changes in heart rate and blood pressure per heart beat.
- Flow through the arterial and following venous subsystems is determined by the organs selected by the body for their need of blood and by the local muscle contraction conditions. Table 7-1 indicates changes in blood supply for various body parts at rest and work.
- Oxygen intake ("uptake") $\dot{V}O_2$ is the volume of oxygen (at defined atmospheric conditions) absorbed per minute from the inspired air. At rest, this volume is about 0.2 L min^{-1}, which can increase at hard exercise up to 30-fold.
- Heart Rate (HR) is the number of ventricular contractions per minute. Its range is usually 60 to 70 beats/min at rest and increases three-fold at strenuous exercise. Pulse rate is the frequency of pressure waves in the arteries and is the same as heart rate in normal healthy individuals.
- Stroke Volume (SV) is the volume of blood ejected from the left heart into the main artery during each ventricular contraction. It is usually 40 to 60 mL at rest and may increase three-fold with hard work.
- Cardiac output, also called (cardiac) minute volume, is the volume of blood injected into the main artery per minute: stroke volume times multiplied with heart rate equals cardiac output. It may increase five- to seven-fold over resting values.

Table 7-1. Blood supply to consuming organs during rest and work (adapted from Astrand and Rodahl 1977).

Consumer	At Rest*	At Heavy Work*
Muscle	15	75
Heart	5	5
Digestive tract	20	3
Liver	15	3
Kidneys	20	3
Skin	5	5
Bone	5	1
Fatty tissues	10	1
Brain	5	5

*estimated percent of the cardiac output which is about 5 L min^{-1} at rest, and 25 L min^{-1} at heavy work.

- Blood Pressure (BP) is the internal pressure in the arteries near the heart, at rest about 70 mmHg during diastole and 120 during systole. These values may double with heavy exercise.

REFERENCES AND FURTHER READING

Astrand, P. 0. and Rodahl, K. 1977 and 1986. *Textbook of Work Physiology*. Second and Third Edition. New York, NY: McGraw-Hill.

Guyton, A. C. 1979. *Physiology of the Human Body*. Fifth Edition. Philadelphia, PA: Saunders.

Chapter 8

THE METABOLIC SYSTEM

OVERVIEW

Over time, the human body maintains a balance (*homeostasis*) between energy input and output. The input is determined by the nutrients, from which chemically stored energy is liberated during the metabolic processes within the body. The output is mostly heat and work, measured in terms of physically useful energy, i.e., energy transmitted to outside objects. The amount of such external work performed strains individuals differently, depending on their physique and training.

The Model: The "Human Energy Machine"

Astrand and Rodahl (1977) used an analogy between the human body and an automobile. In the cylinder of the engine, an explosive combustion of a fuel-air mixture transforms chemically stored energy into physical kinetic energy and heat. The energy moves the pistons of the engine, and gears transfer their motion to the wheels of the car. The engine needs to be cooled to prevent overheating. Waste products are expelled. This whole process can work only in the presence of oxygen and when there is fuel in the tank. In the "human machine," muscle fibers are both cylinders and pistons: bones and joints are the gears. Heat and metabolic byproducts are generated while the muscles work. Nutrients (mostly carbohydrates and fats) are the fuels which must be oxidized to yield energy.

INTRODUCTION

This chapter first provides an overview of metabolism and work and then discusses, in three sections, the process of energy liberation, the assessment of a person's capacity for energy expenditure, and the energy requirements at work. Figure 8-1 shows schematically the main interactions within the body.

Human Metabolism and Work

The term "metabolism" includes all chemical processes in the living body. In a narrower sense, it is often — and here — used to describe the (overall) energy-yielding processes.

The balance between energy input I (via nutrients) and outputs can be expressed by an equation:

$$I = M + W + S \qquad (8\text{-}1)$$

where M is the metabolic energy generated, W the work performed on an outside object, and S the energy storage in the body (negative if loss from it).

The measuring units for energy (work) are Joules (J) or calories (cal) with 4.2 J = 1 cal. (Exactly: 1 J = 1 Nm = 0.2389 cal = 10^7 ergs = 0.948 × 10^{-3} BTU = 0.7376 ft lb.) One uses the kilocalorie, kcal or Cal = 1000 cal, to measure the energy content of foodstuffs. The units for power are 1 kcal hr^{-1} = 1.163 W, with 1 W = 1 Js^{-1}.

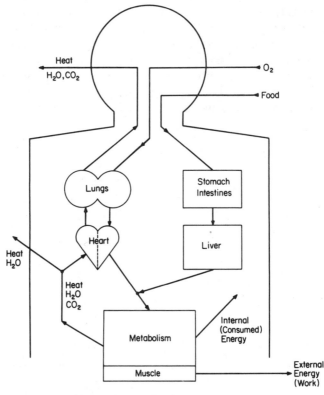

Figure 8-1. Interactions among energy inputs, metabolism, and outputs of the human body.

Assuming for simplicity no change in energy storage and also that no heat is gained from or lost to the environment (see Chapter 9 for details about actual heat exchange) one can simplify the energy balance equation to

$$I = M + W \qquad (8\text{-}2)$$

Human energy efficiency e (work efficiency) is defined as the ratio between work performed and energy input

$$e = \frac{W}{I} \, 100 = \frac{W}{M + W} \, 100 \quad \text{(in percent)} \qquad (8\text{-}3)$$

In everyday activities, only about 5% or less of the energy input is converted into "work," which is energy usefully transmitted to outside objects; highly trained athletes may attain, under favorable circumstances, perhaps 25%. The remainder of the input is, finally converted into heat.

Work (in the physical sense) is done by skeletal muscles (see Chapter 3); they move body segments against external resistances. The muscle is able to convert chemical energy into physical work or energy. From resting, it can increase its energy generation up to fifty-fold. Such enormous variation in metabolic rate not only requires quickly adapting supplies of nutrients and oxygen to the muscle but also generates large amounts of waste products (mostly heat, carbon dioxide, and water) which must be removed. Thus, while performing physical work, the ability to maintain the internal equilibrium of the body is largely dependent on the circulatory and respiratory functions which serve the involved muscles. Among these functions, the control of body temperature is of particular importance. This function interacts with the external environment, particularly with temperature and humidity as discussed in the related chapter in more detail.

The work (output) capability of the human body is dependent on its internal ability to generate energy over various time periods at varying energy levels. The engineer determines the "work" required and how it is to be done and also has much control over the external environment. To arrange for a suitable match between capabilities and demands, the engineer needs to adjust the work to be performed (and the work environment) to the body's energetic capabilities. These human capabilities are determined by the individual's capacity for energy output (physique, training, health); the neuromuscular function characteristics (such as muscle strength, coordination of motion, etc.); and psychological factors (such as motivation).

SECTION 1: ENERGY LIBERATION IN THE HUMAN BODY

Section 1 Overview

Energy is introduced into the body through the mouth in the form of food and drink. The energy carriers are carbohydrates, fats, and proteins. After passing through the stomach, the nutrients are absorbed in the small intestines and waste is separated. The absorbed nutrients are then stored either as glycogen or as (tissue-building) fat.

Energetic Reactions

Energy transformation in living organisms involves chemical reactions that either:

liberate energy, most often as heat. Such reactions are called exergonic (or exother-
mic); or

require energy input; these are called endergonic (or endothermic) reactions.

Generally, breakage of molecular bonds is exergonic, while formation of bonds is
endergonic. Depending on the molecular combinations, bond breakage will yield
different amounts of released energies. Often, reactions do not simply go from the most
complex to the most broken-down state, but achieve the process in steps with intermediate and temporarily incomplete stages.

The Pathways

Energy is supplied to the body as food or drink through the mouth (*ingestion*). Here,
the food is chewed into small particles and mixed with saliva (*mastication*). Saliva
contains mucus which makes the particles of the food stick together in a convenient size
(*bolus*) and lubricates it for passage (*deglutition*) through the pharynx and esophagus to
the stomach.

Churning movements of the stomach help the gastric juice to break up the bolus until it
is fully liquefied (*chyme*). Most of the chemical digestion of foods takes place in the
duodenum, about the first 25 cm of the small intestine (which has a smaller diameter than
the large intestine but is about five times longer). The inner surface of the smaller
intestine has many fingerlike projections (*villi*) which increase the internal surface for
absorption. Blood and lymph capillaries into which digested foods are absorbed are
embedded in these surfaces.

"Digestion" takes place in the stomach and more in the following small intestines. This
is not only a physical reduction of solid to liquefied food, but more importantly changes it
chemically by breaking large complex molecules into smaller ones. These can be
transported through membranes of the body cells and absorbed. The large variety of
chemical reactions on the digested and absorbed foodstuffs is called "assimilation"; it
allows the foodstuffs either to be degraded to release their energy content or to use them
as raw materials for body growth and repair. Solid wastes and undigested food components are egested (*defecated*), having been shaped into a soft solid mass (*feces*).
Nitrogenous wastes produced in the cells are transported by the blood to the kidneys
where they are excreted in the urine.

The Nutrients

Our food consists of various mixtures of organic compounds (foodstuffs) and water,
salts and minerals, vitamins, etc. and of fibrous organic material (primarily cellulose).
This roughage, or bulk, improves mechanical digestion by stretching the walls of the
intestines but does not release energy.

The primary foodstuffs are carbohydrates, fats, and proteins. Their average nutrition-
ally usable energy contents per gram are: 4.2 kcal (18 kJ) for carbohydrate, 9.5 kcal (40
kJ) for fat, and 4.5 kcal (19 kJ) for protein.

The energy value of foodstuffs is measured in a "bomb" calorimeter where food
material is electrically burned so that its content is completely reduced to CO_2, H_2O, and
nitrogen oxides; the heat developed is the measure for the energy content of the food.
The energy content of our daily food (and drink) depends on the mixture of the basic
foodstuffs therein (see Table 8-1). It is of some interest to note that protein oxidized in
the body yields only 4.5 kcal g^{-1}, which is about 75 percent of the energy freed in the

Table 8-1.   Energy content of foods and drinks.   (Based on Pennington and Church 1985, and on data supplied by several fast food chains.).

Item	kcal per serving		kcal per 100 g or 100 ml	kJoule per 100 g or 100 ml
Basic Food Groups				
Dairy Products and Eggs:				
American cheese	(1 oz slice)	106	380	1590
Butter	(1 tbsp)	108	720	3010
Cottage cheese, 2% fat	(1 cup)	203	90	375
Egg (large)	(1)	80	160	670
Ice cream (plain) vanilla	(1 cup)	269	202	845
Yogurt, plain, whole milk	(1 cup)	139	61	256
Fats and Oils:				
Margarine	(1 tbsp)	102	680	2842
Mayonnaise	(1 tbsp)	100	708	2963
Oil, vegetable	(1 tbsp)	120	857	3583
Salad dressing (French)	(1 tbsp)	67	419	1750
Fruits:				
Apples	(1 medium)	81	59	247
Bananas	(1 medium)	105	92	385
Oranges (navel)	(1 medium)	65	46	192
Peaches	(1 medium)	40	46	193
Pears	(1 medium)	98	59	247
Raisins	(1 cup)	450	300	1254
Strawberries	(1 cup)	450	30	126
Vegetables:				
Beans, baked	(1/3 cup)	300	115	482
Beans, lima, cooked	(1/2 cup)	178	111	464
Broccoli, cooked	(1 cup)	39	26	109
Carrots, raw	(1 large)	42	42	176
Celery, raw	(1 stalk)	8	16	67
Corn, sweet, cooked	(14-inch ear)	100	100	418
Lettuce	(1/4 small head)	15	15	63
Mushrooms, raw	(10 small)	28	28	117
Onions, cooked	(1 cup)	58	29	121
Potatoes, white, baked	(1 medium)	95	95	397
Spinach, cooked	(1 cup)	42	23	96
Tomatoes, raw	(1 small)	22	22	92
Grain Products:				
Bisquit (baking powder)	(1)	103	368	1538
Corn flakes	(1 cup)	110	393	1643
Croissant	(1)	109	419	1750
Danish pastry (apple)	(1)	121	327	1367
Oatmeal, cooked	(1 cup)	144	62	259
Rice, white cooked	(1 cup)	205	109	456
Saltines	(4)	52	433	1810
Shredded wheat	(2 bisquits)	166	346	1446
Spaghetti, cooked	(1 cup)	216	148	619
White bread	(1 slice)	64	267	1116
Wholewheat bread	(1 slice)	61	244	1020

(continued)

Table 8-1. Energy content of foods and drinks. (Based on Pennington and Church 1985, and on data supplied by several fast food chains.). (CONTINUED)

Item	kcal per serving		kcal per 100 g or 100 ml	kJoule per 100 g or 100 ml
Basic Food Groups (continued)				
Meat, Poultry, Fish:				
Bacon, fried	(1 slice)	35	583	2437
Chicken, roasted w/o skin	(4 oz)	217	190	794
Hamburger, cooked	(4 oz patty)	300	264	1103
Hot dog, beef	(1)	145	322	1346
Roast beef, lean	(4 oz)	190	168	702
Steak, cooked	(4 oz)	240	210	878
Turkey, light meat	(4 oz)	179	187	656
Halibut, cooked	(4 oz)	195	171	715
Red snapper, raw	(4 oz)	106	93	390
Salmon, cooked	(4 oz)	208	182	761
Shrimp, raw	(4 oz)	104	91	380
Swordfish, cooked	(4 oz)	199	174	727
Trout, brook, cooked	(4 oz)	224	196	819
Tuna, canned in oil	(4 oz)	238	210	878
canned in water	(4 oz)	126	111	464
Other:				
Sugar	(1 cup)	770	385	1610
Honey	(1 tbsp)	61	305	1295
Jam	(1 tbsp)	55	275	1150
Snack Foods				
Almonds, slivered	(1/4 cup)	170	515	2155
Brownies	(1)	130	406	1697
Chocolate chip cookies	(4)	184	460	1923
Chocolate covered peanuts	(1 oz)	153	546	2282
Cupcakes, chocolate icing	(1)	130	360	1510
Hard candy	(1 oz)	110	390	1630
Milk chocolate candy	(1 oz)	157	533	2228
Oatmeal cookies w/raisins	(40)	244	470	1965
Potato chips	(10)	115	575	2405
Pretzels	(1 oz)	111	396	1655
Roasted peanuts	(1 cup)	840	585	2445
Vanilla wafers	(4)	68	463	1938
Popsicle	(1 bar)	65	74	309
Beverages				
Cola	(12 fl oz)	145	41	172
Coffee	(12 fl oz)	4	2	8
Diet cola	(12 fl oz)	1	0	0
Hot chocolate	(12 fl oz)	330	87	364
Tea	(12 fl oz)	0	0	0
Milk, whole	(12 fl oz)	200	63	264
Milk, low fat	(12 fl oz)	162	50	207
Milkshakes	(12 fl oz)	350-380	120-130	510-545
Orange or apple juice	(12 fl oz)	160	45	188
Gatorade	(12 fl oz)	60	17	71

(continued)

Table 8-1. Energy content of foods and drinks. (Based on Pennington and
Church 1985, and on data supplied by several fast food chains.). (CONTINUED)

Item	kcal per serving		kcal per 100 g or 100 ml	kJoule per 100 g or 100 ml
Beverages (continued)				
Beer: Regular	(12 fl oz)	144-162	41-46	172-192
Light	(12 fl oz)	96-135	27-38	113-159
Wine (red and white)	(3-1/2 fl oz)	80	78	326
Liquor (bourbon, gin,				
scotch, vodka):	jigger of			
80 proof	(1-1/2 fl oz)	95	225	940
90 proof	(1-1/2 fl oz)	110	260	1085
Brandy	(1-1/2 fl oz)	120	285	1195
Mixed drinks	(serving)	110-320		
"Fast Foods"				
"Single" hamburger	(100-125 g)	260-340	260-270	1090-1130
"Specialty" hamburger				
(e.g., Deluxe, Double,				
Big Mac, Whopper)	(200-280 g)	550-670	240-275	1005-1150
Sandwiches: Chicken	(130-210 g)	320-690	250-330	1045-1380
Fish	(140-205 g)	435-540	265-310	1110-1300
Roast beef	(150 g)	320-350	215-235	900-985
Fried chicken	(40-70 g)	105-200	265-285	1110-1190
French fries, regular	(70 g)	210	300	1255
Bean burrito		345		
Beef burrito		470		
Tostada (bean)		180		
Beef tostada		290		
Taco		190		
Cheese pizza, reg. crust	(1/4 of	325	231	970
Pork with mushroom pizza	12" diam.	380-500		
"Super supreme" pizza	pizza)	520-590		

Volume units: 16 tbsp = 1 cup = 8 fl oz = 237 ml
Mass units: 1 oz = 28.4 g

(continued)

Table 8-1. Energy content of foods and drinks. (Based on Pennington and Church 1985, and on data supplied by several fast food chains.). (CONCLUDED)

Prepared Meals

Breakfast No. 1:
      Fried eggs, two
      Bacon, 3 slices
      Hash brown potatoes
      Wholewheat toast, 2 slices, with butter   Total: about 700 kcal

Breakfast No. 2:
      Sweet roll, one   Total: about 200 kcal

Breakfast No. 3:
      Corn flakes, 1 cup with 1/2 cup milk   Total: about 200 kcal

Lunch No. 1:
      Lean roast beef on French bread, 2 slides
        w/ 1 tbsp of mayonnaise
      Chips, ten   Total: about 550 kcal

Lunch No. 2:
      Batter-fried fish, 3 oz
      French fries, 10 large   Total: about 350 kcal

Lunch No. 3:
      Salad (1/2 head of lettuce, 1 tomato, croutons)
        w/2 tbsp of Thousand Island dressing   Total: about 240 kcal

Dinner No. 1:
      Lean roast beef, 4 oz
      Mashed potatoes, 1 cup
      Bread, 1 slice, w/butter
      Peas, 1/2 cup   Total: about 650 kcal

Dinner No. 2:
      Spaghetti and meatballs, 1 cup
      Salad (lettuce, tomato, croutons)
        w/Italian dressing   Total: about 570 kcal

Dinner No. 3:
      Chili w/beans, 1 cup
      Corn muffins, two, w/butter   Total: about 565 kcal

calorimeter. Hence, energy liberation from protein in the body is less efficient than from fats and carbohydrates.

Carbohydrates range from small to rather large molecules, and most are composed of only the three chemical elements C, O, and H. (The ratio of H to O usually is 2 to 1, as in water, hence the name carbohydrate, meaning watered carbon). Carbohydrates exist as simple sugars (*monosaccharides*), double sugars (*disaccharides*), and *polysaccharides* which are a large number of monosaccharides joined into chains. The most common natural polysaccharides are plant starch, glycogen, and cellulose. Carbohydrates are digested by breaking the bonds between monosaccharides so that the compounds are reduced to simple sugars, such as glucose which can be absorbed.

Fat is called a triglyceride because a fat molecule is formed by joining one glycerol nucleus to three fatty acid radicals. Unsaturated fat has double bonds between adjacent carbon atoms in one or more of the fatty acid chains; hence the compound is not saturated with all the hydrogen atoms it could accommodate. Some of the available bonds are in fact not occupied by hydrogen atoms but increase the number of linkages between carbon atoms themselves. The more unsaturated a fat is, the more it is liquid, an oil. Most plant fats are polyunsaturated while most animal fats are saturated and hence solid. (Diet with a high content of saturated fat is medically suspect since it has been linked with high blood pressure.)

Digestion of fat means the breakage of bonds which link the glycerol residue to the three fatty acid residues. Fats are digested primarily in the small intestine. (Glycerol and fatty acid molecules are small enough to cross cell membranes and hence can be absorbed.) The bloodstream can transport only water-soluble materials such as glycerol, but many fatty acids are water repellent: these are absorbed into the lymph vessels. Some fat is absorbed as triglycerides (neutral fat) in the adipose tissue.

Proteins are chains of amino acids joined together by peptide bonds. Many such different bonds exist, and hence proteins come in a large variety of types and sizes. A special group of proteins functions as enzymes, i.e., organic catalysts which speed up chemical reactions between other molecules without being consumed themselves. Enzymes secreted by the stomach and the pancreas are components of the digestive juices, which play an important role in the chemical digestion. Proteins as foodstuffs are partly digested in the stomach and then, in steps, in the small intestine.

Absorption and Assimilation of Foodstuffs

In the stomach, primarily water, salts, certain drugs and alcohol are absorbed. Most of the digested foodstuffs are assimilated in the small intestine. Here, all digested foodstuffs enter the blood capillaries or the lymphatic system. The blood capillaries drain eventually into the hepatic portal vein which carries blood to the liver. This vein also receives inputs from the stomach, pancreas, gall bladder, and spleen. Lymph flows from the intestinal walls through the thoracic duct to the left subclavian vein. Here the lymph enters the blood stream and becomes part of the blood plasma. The liver receives blood not only from the portal vein but also from the hepatic artery.

Liver cells remove digestion products from the blood and store or metabolize them. Particularly important is the formation of glycogen, the storage form of carbohydrate, from glucose. (This can also be done at muscle cells.) Fat is being synthesized in the liver from glucose (and from amino acids derived from proteins) which then serves as energy storage. Thus, the liver controls much of the fat utilization of the body.

Assimilation of foodstuffs can be in two complementary processes. In *anabolism* or constructive metabolism, small molecules are assembled into larger ones by chemical reactions requiring an input of energy (endergonic metabolism). This needed energy is supplied by catabolism, or destructive metabolism, in which organic molecules are broken down releasing their internal bond energies (exergonic metabolism).

Aerobic Metabolism

The monosaccharides produced in the digestion of food carbohydrates are principally glucose (80%), fructose and galactose; the latter two monosaccharides are also quickly converted to glucose. Glucose can be oxidized according to the formula:

$$C_6H_{12}O_6 + 6O_2 = 6CO_2 + 6H_2O + \text{Energy} \qquad (8\text{-}4)$$

This means that one molecule of glucose combines with six molecules of oxygen and results in six molecules each of carbon dioxide and water, while energy (about 690 kcal/mole) is released.

Chemically, oxidation is defined as a loss of electrons from an atom or a molecule (while reduction is defined as a gain of electrons). In such reactions, electrons are carried in the form of hydrogen atoms, and the oxidized compound is de-hydrogenated. In human metabolism, organic fuels (glucose, fats, and occasionally proteins) constitute the major electron donors, while oxygen is the final electron acceptor (oxidant) of the fuel.

Anaerobic Metabolism

Another way of oxidation is the breakdown of glucose and glycogen molecules into several fragments and these fragments becoming oxidized by each other. In this case, energy yield is anaerobic, and the processes are called glycolysis and glycogenolysis, respectively.

Glucose (and fat) catabolism takes place in a number of small steps, i.e., in sequential biochemical reactions in which intermediary metabolites are produced. Two phases are of particular interest. The first is anaerobic glycolysis. Here, the 6-carbon compound glucose is broken into two 3-carbon molecular fragments each of which naturally becomes a molecule of the 3-carbon compound pyruvic acid. The second phase is aerobic, meaning that oxygen is required. This is a series of self-renewing reactions known as the Krebs (citric acid) Cycle (see Figure 8-2). Here, hydrogen atoms are separated in pairs from the intermediary metabolites, the first of which is pyruvic acid. After the pyruviates pass through the Krebs Cycle, glucose is completely metabolized and six CO_2 molecules and six H_2O molecules are produced, as per equation 8-4. The removal of hydrogen atoms from the intermediary metabolites (particular to the Krebs Cycle) is called dehydrogenation.

The Energy Carriers

Glucose is the primary, most easily accessible and most metabolized energy carrier for the human body. However, fats (mostly in adipose tissue) account for most of the stored fuel reserves. For energy release, neutral fat first needs to be split into glycerol and fatty acids. Similar to the (much simpler) breakage of glucose bonds, glycerol and fatty acids are converted to 3-carbon and 2-carbon intermediary metabolites and enter the Krebs Cycle. The 3-carbon mebabolite enters the glucose pathway while the 2-carbon compound is oxidized to CO_2 and H_2O. The energy yield of fats per volume is approximately 2340 kcal/ mole, i.e. nearly 3.5 times that of carbohydrates.

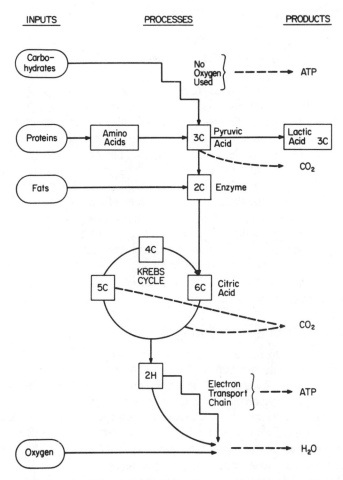

INPUTS   PROCESSES   PRODUCTS

Figure 8-2. Breakdown of Foodstuffs (modified from Astrand and Rodahl 1977), showing the number of carbon atoms in the steps.

Almost all the results of protein digestion are absorbed in the form of amino acids. Under normal circumstances the body does not use proteins as a source of energy since their catabolism usually involves the death of cells, protein being part of the protoplasm of living cells. Catabolism of proteins does occur, however, in fasting, malnutrition, starvation, and certain illnesses.

Energy Storage and Release

Living cells store "quick-release" energy in the molecular compound *adenosine triphosphate*, ATP. Its phosphate bonds can be broken down easily by hydrolysis, thus providing quick energy for driving the endergonic process.

$$ATP + H_2O \rightarrow ADP + energy \text{ (output)} \tag{8-5a}$$

ATP is an "intracellular vehicle of chemical energy"; it transfers its energy by donation of its high-energy phosphate group to processes that require energy. But the ATP supply needs to be replenished constantly. This is done through *creatine phosphate*, CP, which transfers a phosphate molecule to *adenosine diphosphate*, ADP. Energy must be supplied for this reaction:

$$ADP + CP + energy\ (input) \rightarrow ATP + H_2O \qquad (8\text{-}5b)$$

The cycle of converting ATP into ADP, releasing energy for muscle action and of re-converting ADP into ATP, does not require the presence of oxygen.

While ATP can provide quick energy anaerobically for a few seconds, resynthesis of ATP is necessary for continuous operation. The energy source for this process is the breakdown of complex molecules (provided by the absorbed foodstuffs) to simpler ones, ultimately to CO_2 and H_2O. Hence, glucose, fats, and proteins provide the ultimate source of energy, keeping the ATP—ADP process going.

As discussed, the use of foodstuffs does *not* require at *every* moment, and in *every* step, the immediate participation of oxygen; although, in the final balance, oxygen is required. Most of the single-step oxidations in the biological reactions are in fact anaerobic, dehydrogenations where hydrogen atoms are carried through several reactions. Thus, the muscle cell can work (during many breakdown steps of the fuel) anaerobically, that is in the absence of oxygen; but finally, oxygen must be provided. Thus, overall, sustained energy use is aerobic, i.e., it requires oxygen.

Liberation and Use of Energy During Work

As just discussed, the liberated energy stems from the breakage of fatty acid and glycerol bonds and of glucose bonds. The combustion of the fat derivatives is strictly aerobic. Glucose breakdown is first anaerobic followed by an aerobic phase but needs oxygen for complete metabolism. Glycogen stores near muscles are depleted much more quickly when the muscles must work anaerobically than when able to work aerobically.

Usually, at rest and during moderate work, the oxygen supply is sufficient and, hence, the energy metabolism is essentially aerobic. This leads to high ATP concentration and low ADP content. To meet intermediate work (energy) demands, the breakdown of glucose speeds up. At a critical intensity, the oxygen-transporting system cannot provide enough oxygen to the cells and pyruvate is transformed into lactic acid instead of going through the Krebs Cycle. During continued intermediate work, the lactate developed may be reconverted to glycogen (in the liver, perhaps even at the muscle — Astrand and Rodahl, 1986) if aerobic conditions exist. With high work intensity, increasingly more parts of the metabolic processes are anaerobic, and lactic acid buildup increases which may eventually require cessation of the muscular work.

Looking more minutely at the processes of energy liberation during the first 45 to 90 seconds of exercise, one finds that (when the work intensity is below the maximum possible but above about 50%) some anaerobic energy metabolism takes place while the oxygen supply is not yet adjusted to meet the demand.

Table 8-2 shows schematically the contributions of aerobic and anaerobic energy liberation during short and long maximal efforts — of course, the intensity magnitude of the effort is higher in a short burst of output than in sustained exertion.

During maximal work of short duration (up to one minute or so) the energy available depends on ATP splitting while during prolonged heavy exercise (longer than one hour), the maximal work output depends on the oxidation of fatty acids and glycogen. The most complex process in terms of fuel conversion is during hard work that lasts from about 1 to 10 minutes. At the start of such rigorous work, utilization of ATP and phosphocreatine is predominant. Then anaerobic conversion of glucose to lactate takes over increasingly.

Table 8-2. Contributions of anaerobic and aerobic energy liberation during maximal efforts of various durations (schematically from Astrand and Rodahl 1977).

Energy Released	Duration of the Greatest Possible Effort either				
	10 s	or 1 min	or 10 min	or 1 hr	or 2 hr
by Anaerobic Processes					
in kJ	100	170	150	80	65
in %	85	65 to 70	10 to 15	2	1
by Aerobic Processes					
in kJ	20	80	1000	5500	10000
in %	15	30 to 35	85 to 90	98	99
Total in kJ	120	250	1150	5580	10065
Primary Energy Source	ATP Splitting	CP	Glucose Glycolysis	Glucose and Fatty Acids Krebs Cycle	
Process	Anaerobic	Mixed		Aerobic	

During the final phase, the oxidation of glucose and eventually of fatty acids predominates.

Summary of Section 1 "Energy Liberation in the Human Body"

- The metabolic breakdown of foodstuffs releases energy which is, to a small part, transmitted as useful mechanical energy to an outside object (the person "works") but mostly transformed into heat.
- While, overall, the human metabolic process liberates energy (exergonic), reactions within this process require energy (endergonic).
- The primary energy suppliers are carbohydrates and fats. Protein provides enzymes as catalysts in the digestion, and it is used for tissue re-building but normally not for energy.
- The oxidation of foodstuffs provides the overall energy source for the metabolic processes while the breakdown of ATP into ADP is the immediate energy source.
- The breakdown of glucose is in aerobic as well as anaerobic steps, with the latter often resulting in the build-up of lactic acid — a metabolite that can force cessation of the effort if not re-synthesized. Glucose provides most of the energy for maximal efforts lasting less than one hour.
- The breakdown of fatty acids is aerobic, i.e., it needs oxygen. It provides much of the energy for heavy physical work lasting hours.

SECTION 2: ASSESSMENT OF ENERGY EXPENDITURES

Section 2 Overview

To match a person's work capacity with the job requirements, one needs to know the individual's energetic capacity, and how much a given job demands of this capacity. This section primarily addresses the first topic, the next section deals with job demands.

Among the currently used procedures to assess internal metabolic capacities, four techniques are predominant:

  diet and weight observation;
  direct calorimetry;
  indirect calorimetry; and
  subjective rating of perceived strain.

Introduction

The following text discusses several well established techniques to measure human energy exchange. However, "insight and caution" are in order here so that the results of these techniques can be properly understood and applied.

Problem #1. Human energy balance must be observed over a sufficiently long time period so that delays and advances in energy intake, conversion, and output "average out."   In the previous equations, energy storage was (for convenience) excluded. However, imbalances between energy intake and output of hundreds of kilocalories over a day are quite common. Most of the energy store in a healthy adult (fat, glycogen, and eventually protein) is in the form of adipose tissue of the magnitude of 100,000 kcal. A change in these components would result in a change in body weight, which is easily observable. (Many data on energy consumption are given in relation to body weight.) However, the seemingly simple relationship between weight and energy store is disturbed by water which is a large and rather labile component of body weight but contributes nothing to the energy stores (Garrow 1980).

Problem #2. Assessment of the amount of total body fat or conversely of lean body mass, is usually done by measuring skinfold thickness and inserting the results into various equations which should yield values for body fat or lean mass (Roebuck, Kroemer, and Thompson 1975). Altogether, this is not very accurate.

Problem #3. The fact and conditions of measuring the energy exchanges may influence the measured results: putting on and carrying the measuring apparatus, hindrance of motions and breathing, and being observed in general may lead to energetic and circulatory processes different from those under normal circumstances.

Problem #4. Inaccuracies made in the measurements and/or of inadequacies in the experimental procedure may degrade the results. As discussed in more detail later, energy output measurements should be taken during the "steady state" of work, after the functions have reached stability, and well before a rest period. This may be rather difficult to do, for example, if the time available for the measurement is very short or if work intensity or the environment change, thus not allowing a steady state to develop. In fact there may be no steady state in short term efforts or in efforts near or beyond one's capacity.

Problem #5. For data, one needs a "baseline," or reference. One would like to start with the basal metabolism (or heart rate) and observe the increase from this value. However, a true reliable basal condition is quite difficult to achieve in a subject; hence, observed differences from that basal value are subject to unaccounted variations in the baseline, and of course also in the values measured under stress. This is of particular concern for light efforts: say, the measuring accuracy for energy at both rest and work is in the neighborhood of 5%, and assume a "true" energy expenditure increase at work of 10% over rest. In this case, one might actually observe differences in expenditures between 0% and 20%; the first value is obviously nonsense, and the second double the "truth." (This may in fact explain some of the (percent-wise) enormous variations found in different texts concerning the energy requirements of light work.)

Problem #6. In experimental design, one needs to clearly distinguish between the independent variable (the effects of which one would want to study in the experiment) and the dependent variable (which provides that information). In one case, we wish to know about the efforts one must undergo to do work of different "heavinesses," i.e., the effects of various work loads should be observed. The dependent measure would be the oxygen consumption, heart rate, or another related strain indicator of the human. For this one must assume that the human subject is "standardized," meaning that the reactions are "normal," known, and reliable. In the other case, information is desired about the individual capacity of the subject in relation to a known work load. The same reactions of the human as before (e.g. metabolic or circulatory responses) may be used as dependent measures. In this case the work load must be "standardized." In laboratory testing, these two cases can be clearly distinguished and controlled: standardized bicycle ergometer or treadmill loads allow judgments about a given subject's capacity related to such work loads (the second case); this information "standardizes" the subject (with respect to the reactions to such work loads) who then can be used to assess how strainful it is to do certain unknown work (the first case).

Unfortunately, simple and well-controlled laboratory conditions do not exist in the work world. Here, one often has to assess the "heaviness" of the work through the reactions of subjects that have not been "standardized" before. Many of the data describing the energy expenditures, or heart rates, associated with certain jobs or professions or occupations are highly suspect since there is little or no assurance that the field data on energy consumption or heart rate were indeed measured on "standard subjects." Therefore, one has to use such information with great caution: the wide diversity in existing data may be partly explained by variability in the subjects employed and of course by variations in their tasks.

Diet and Weight Observation

In terms of energy, what is inputed into the body as nutrients, solid or fluid, must be outputed either in terms of energy internally consumed to maintain the body (finally converted to heat) or as external work performed. This "energy balance" assumes that there is neither energy storage (when the person gets heavier, particularly fatter) nor use of stored energies (the person getting slimmer). Assessment of internal energetic processes through diet and weight observation provides reliable information about the long-term energy exchange but requires rather lengthy observation periods (weeks or more) and strict complete control over food intake, water balances, and energy output. This is acceptable and useful for laboratory tests but rather impractical for everyday applications.

Direct Calorimetry

Since energy taken in by the body is finally transformed into heat (if no external work is done), one can enclose the human body with an energy-tight chamber which allows measurement of all heat generated and transmitted by conduction, convection, radiation, and evaporation. There are several requirements for this procedure: the room must be small, which reduces the capability to perform work; air (which carries energy) must be supplied to and taken from the chamber; since the set-up is not massless, energy is exchanged with and stored in equipment, walls, etc. Hence, direct calorimetry must be performed in special laboratories and is not of much practical interest. Therefore, most current calorimetric techniques rely on indirect calorimetry.

Indirect Calorimetry

While performing work, the oxygen consumption (and CO_2 release) is a measure of metabolic energy production which can be assessed with a variety of instruments. They all rely on the principle that the difference in O_2 (and CO_2) content between the exhaled and inhaled air indicates the oxygen absorbed (or carbon dioxide released) in the lungs. Given sufficiently long observation periods (more than 5 minutes), this is a reliable assessment of the metabolic processes. Assuming an overall "average" caloric value of oxygen of 5 kcal/L O_2, one can calculate the energy conversion occuring in the body from the volume of oxygen consumed.

A more exact assessment of the nutrients actually metabolized can be made by using the respiratory exchange quotient, RQ, which compares the carbon dioxide expired to the oxygen consumed. One gram of carbohydrate needs 0.83 liter oxygen to be metabolized and releases the same volume of carbon dioxide (see equation 8-4). Hence, the RQ is one (unit). The energy released is 18 kJ per gram, 21.2 kJ or 5.05 kcal/L O_2. Table 8-3 shows these relationships also for fat and protein conversion.

Most medical and physiological assessments of human energetic capabilities currently rely on various techniques of measuring oxygen consumption. To compare a persons' capacities with each other, one relies on standardized tests with normalized external work, mostly using bicycle ergometers, treadmills, or steps. (Selection of this equipment was and is not based so much on theoretical considerations but rather on availability and ease of use.)

Bicycles, treadmills, and steps stress certain body parts and body functions. Bicycling requires predominantly use of the leg muscles. Since the legs constitute both in their mass and their musculature rather large components of the human body, their extensive exercising in a bicycle test also strains pulmonary, circulatory, and metabolic functions of the body. However, a person who is particularly strong in the upper body but not well trained in the use of legs would show different strain reactions in bicycle ergonometry than, say, a bicycle racer.

Table 8-3. Oxygen needed, RQ, and energy released in nutrient metabolism.

	O_2 Consumed $(L\ g^{-1})$	$RQ = \dfrac{Vol\ CO_2}{Vol\ O_2}$	$kJ\ g^{-1}$	$kJ\ L^{-1}O_2$	$kcal\ L^{-1}O_2$
Carbohydrate	0.83	1.00	18	21.2	5.05
Fat	2.02	0.71	40	19.7	4.69
Protein	0.79	0.80	19	18.9	4.49
Average*	NA	NA	NA	21	5

*Assuming the construct of a "normal" adult on a "normal" diet doing "normal" work.

The treadmill also stresses primarily lower body capabilities, but in contrast to bicycling the whole body weight must be supported and propelled by the feet. If the treadmill is inclined, the body must also be lifted. Hence, this test strains the body in a somewhat more complete manner than bicycling but still lets trunk and arm capabilities out of consideration. Furthermore, it requires somewhat more bulky equipment than a stationary ergometer bicycle.

Another method of standardizing stress conditions is to have the subject step onto a platform, step down, step up, etc., for the duration of the test. This "step test" technique requires the simplest possible equipment and stresses body functions in a fashion somewhat similar to running on a treadmill, however by forcing the subject mostly to elevate the total body weight instead of moving it forward. Again, muscular capabilities of the upper body are not tested at all, but a person heavier in weight and with shorter legs than another subject would show larger energy consumption.

Obviously, these three techniques differ in such that they involve different body masses and muscles, and hence strain local (muscular) or central (pulmonary, circulatory, metabolic, etc.) capabilities in somewhat different ways. Improvements on the equipment providing the external stress have been proposed in many different ways. Two examples: a ladder-mill on which one climbs using both arms and legs stresses nearly all body functions; an ergometer bicycle with simultaneous cranking or pushing and pulling with the hands; both strain the body rather completely.

Whichever method to develop the external stress is used, the reactions of the individual body in terms of oxygen consumption and hence energy metabolism are collected and compared to the data gained on other subjects. (Brief descriptions of the techniques to measure oxygen consumption are in Appendix A to this chapter.) Statements can be made regarding the given subject's capabilities with respect to sample or population data.

Indirect calorimetry by heart rate. There is close interaction between the circulatory and metabolic processes. Nutrients and oxygen must be brought to the muscle or other metabolizing organs and metabolic byproducts removed from it for proper functioning. Therefore, heart rate (as a primary indicator of circulatory functions) and oxygen consumption (representing the metabolic conversion taking place in the body) have a linear and reliable relationship in the range between light and heavy work, shown in Figure 8-3. (However, this relationship may change within a person with training, and it is different between persons.) Given this relationship, one often can simply substitute heart rate measurements for measurement of metabolic processes, particularly for O_2 assessment. This is a very attractive shortcut since heart rate measurements can be performed easily.

The simplest method for heart rate counting is to palpate an artery, often in the wrist or perhaps in the neck, or to listen to the sound of the beating heart. All the measurer needs to do is count the number of heart beats over a given period of time (such as 15 seconds) and from this calculate an average heart rate per minute. More refined techniques utilize various plethysmographic methods, which rely on deformations of tissue due to changes in filling of the imbedded blood vessels. These methods range from measuring mechanically the change in volume of tissues, for example in a finger, to using photoelectric techniques that react to changes in transmissibility of light depending on the blood filling, such as of the ear lobule. Other techniques rely on electric signals associated with the pumping actions of the heart (ECG), sensed by electrodes placed on the chest.

These techniques are limited in their reliability of assessing metabolic processes primarily by the intra- and inter-individual relationships between circulatory and metabolic functions. Statistically speaking, the regression line (shown in Figure 8-3)

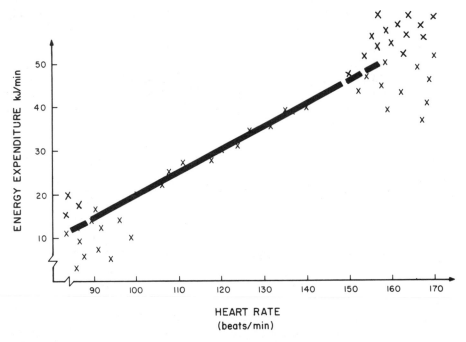

Figure 8-3. Scheme of the relationships between oxygen uptake (expressed as energy expenditure) and heart rate.

relating heart rate to oxygen uptake (enery production) is different in slope and intersect from person to person and from task to task. In addition, the scatter of the data around the regression line, indicated by the coefficient of correlation, is also variable. The correlation is certainly low at light loads where the heart rate is barely elevated and circulatory functions can be influenced easily by psychological events (excitement, fear, etc.) which may be completely independent of the task proper. With very heavy work, the O_2—HR relation may also fall apart, for example when cardiovascular capacities may be exhausted before metabolic or muscular limits are reached. Presence of heat load also influences the O_2—HR relationship.

Use of heart rate has a major advantage over oxygen consumption as indicator of metabolic processes: it responds faster to work demands, hence indicates more easily quick changes in body functions due to changes in work requirements.

Indirect calorimetry by subjective rating of perceived effort. The human is able to perceive the strain generated in the body by a given work task and to make absolute and relative judgments about this perceived effort: one can assess and rate the relationships between the physical stimulus (the work performed) and its perceived sensation (Pandolph 1983). This correlation between the psychologically perceived intensity of physically described stimuli has been used probably as long as people exist to express one's preference of one type of work over another. In the 1970s, Borg developed formal techniques to rate the perceived exertion (RPE) associated with different kinds of efforts. For example, one can assess the perceived effort using a nominal scale ranging from "very light" to "hard." Such a verbally anchored scale can be used to "measure" the strain subjectively perceived while performing standardized work, allowing (similar to the methods previously described) a relative assessment of a person's capability to perform stressful work. (Examples of Borg Scales are given in Appendix B to this chapter.)

Summary of Section 2 "Assessment of Energy Expenditures"

Among the techniques to assess metabolic processes, observation of diet and weight, and direct calorimetry are feasible under well controlled laboratory conditions. More practical are the measurement of oxygen uptake and, relatedly, of heart rate. These measurements can be performed over relatively short periods of time with commercially available equipment. (Techniques are discussed in detail, e.g. by Kinney 1980; Mellerowicz and Smodlaka 1981; Stegemann 1984; Eastman Kodak 1983, 1986) Their primary utility lies in the assessment of a given person's metabolic/circulatory capabilities in response to standardized work, often measured using ergometers of the bicycle or treadmill type. Another common technique is Borg's rating of subjectively perceived effort while performing certain work.

SECTION 3: ENERGY REQUIREMENTS AT WORK

Section 3 Overview

This section discusses the demands posed on the body metabolism by different activities.

Procedures to Categorize Metabolic Requirements

Assuming that, without energy storage, all metabolically developed energy is transformed into either externally performed work, or (finally) heat, one customarily assesses the energy requirements imposed on the body within categorized activity levels.

Basal metabolism. A minimal amount of energy is needed to keep the body functioning even if no activities are done at all. This basic metabolism is measured under strict conditions, usually including fasting for 12 hours, protein intake restriction for at least two days, with complete physical rest in a neutral ambient temperature. Under these conditions, the basal metabolic values depend primarily on age, gender, height, and weight, with the last two variables occasionally replaced by body surface area. Altogether there is relatively little inter-individual variation, hence a commonly accepted figure is 1 kcal (4.2 kJ) kg^{-1} hr^{-1} or 4.9 kJ min^{-1} for a person of 70 kg.

Resting metabolism. The highly controlled conditions under which basal metabolism is measured (mostly for medical purposes) are rather difficult to accomplish for practical applications. One usually measures the metabolism before the working day, with the subject as well at rest as possible. Depending on the given conditions, resting metabolism is 10% to 15% higher than basal metabolism.

Work metabolism. The increase in metabolism from resting to working is called work metabolism. This increase above resting level represents the amount of energy needed to perform the work.

At the start of physical work, oxygen uptake initially follows the demand sluggishly. As Figure 8-4 shows, after a slow onset oxygen intake rises rapidly and then slowly approaches the level at which the oxygen requirements of the body are met. During the first minutes of physical work, there is a discrepancy between oxygen demand and available oxygen. (During this time, the energy yield is largely anaerobic.) This *oxygen*

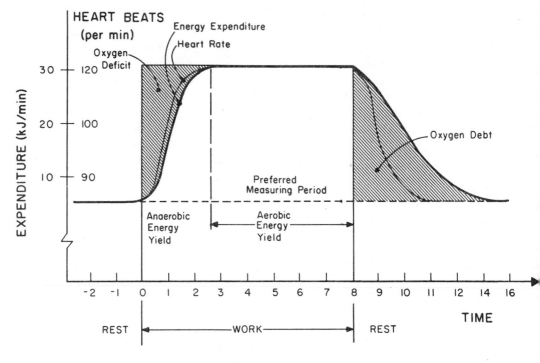

Figure 8-4. Scheme of energy liberation, energy expenditure, and heart rate at steady state work.

deficit must be repaid at some time, usually during rest after work. The amount of this deficit depends on the kind of work performed and on the person.

After stopping the work, the oxygen demand falls again to a resting level, fast at first and then leveling off. During this time, the depleted ATP stores are refilled, but elevated tissue temperature and epinephrine concentration, increased cardiac and respiratory functions, and other phenomena require (as Astrand and Rodahl put it) that "the body pay 100% interest on the oxygen borrowed from the anaerobic bank": the *oxygen debt* repaid is approximately twice as large as the oxygen deficit incurred. Of course, given the close interaction between the circulatory and the metabolic systems, heart rate reacts similarly; but as it increases faster at the start of work than oxygen uptake it also falls back more quickly to the resting level.

An explanation for the sluggish responses of oxygen intake to the start of external physical work lies in the time it takes to increase the flow of oxygen-rich blood to the internal consumers. Some oxygen is stored in the muscles bound to myoglobin and in the blood profusing the muscles. At the beginning of muscular labor, this stored oxygen does not suffice and anaerobic processes take place to release energy for the muscular work.

This results in the production of lactic acid. Insufficient blood flow also leads to excessive build-up of potassium (Kahn and Monod 1989). Lactic acid and potassium accumulation are believed to be the primary reasons for "muscle fatigue" forcing the stoppage of muscle work.

If the workload does not exceed about 50% of the worker's maximal oxygen uptake, then oxygen uptake, heart rate, and cardiac output can finally achieve the required supply level and can stay on this level. This condition of stabilized functions at work is called "steady state." Obviously, a well-trained person can attain this equilibrium between demands and supply even at a relative high workload, while an ill-trained person would be unable to attain a steady state at this requirement level but could be in equilibrium at lower demand.

If the energetic work demands exceed about half the person's maximal O_2 uptake capacity, anaerobic energy-yielding metabolic processes play increasing roles. In particular, this results in high lactic acid production. The length of time during which a person performs this work depends on the subject's motivation and the will to overcome the feeling of "fatigue," which usually coincides with depletion of glycogen deposits in the working muscles, drop in blood glucose, and increase in blood lactate. However, the processes involved are not fully understood, and highly motivated subjects may maintain work that requires very high oxygen uptake for five or six minutes, while other persons feel that they must stop after just two or three minutes of effort.

When severe exercise brings about a continuously growing oxygen deficit and an increase in lactate content of the blood because of anaerobic metabolic processes, a balance between demands and supply cannot be achieved; no "steady state" exists, the work requirements exceed capacity levels (see Figure 8-5). The resulting fatigue can be counteracted by the insertion of rest periods. Given the same ratio of "total resting time" to "total working time," many short rest periods have more "recovery value" than a few long rest periods.

It is apparent that calorimetric or heart rate measurements should not be done during the rising phases of oxygen intake or heart rate but that these measurements be performed during the steady state period where the physiological functions have achieved equilibrium. Measurements during the first few minutes are not indicative of the demand-and-supply functions between work and body. Likewise, measurement after the cessation of the work is not a direct indicator of the severity of the preceding work demands and of their physiological responses; however, under certain highly controlled conditions, counting the "recovery pulse" has been suggested. Here, the heart rate is monitored from the exact moment where the work was stopped throughout predetermined intervals of the recovery period.

Techniques To Estimate Energy Requirements

There are various ways to describe work or exercise activities: e.g., by the amount of work to be done with the arms or with the legs, by the involvement of the trunk, by the total amount of external work being done, each over the periods of time during which these activities occur, etc. Tables of energy expenditures have been compiled for body postures and for many professional or athletic activities — see Tables 8-4, 8-5, and 8-6.

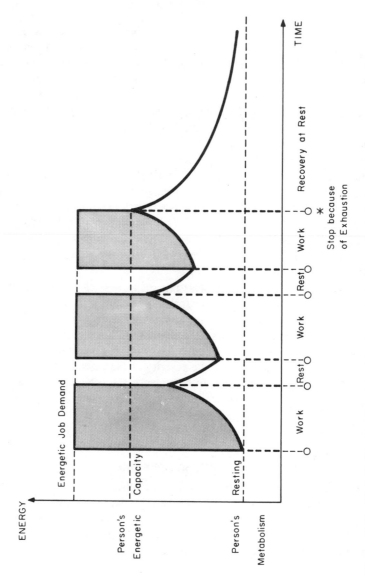

Figure 8-5. Metabolic reactions to the attempt of doing work that exceeds one's capacity even with interspersed rest periods.

Table 8-4.   Energy consumption (to be added to basal metabolism) at various activities (adapted from Astrand and Rodahl 1977, Guyton 1979, Rohmert and Rutenfranz 1983, Stegemann 1984).

	kJ min^{-1}
Lying, Sitting, Standing	
Resting while lying	0.2
Resting while sitting	0.4
Sitting with light work	2.5
Standing still and relaxed	2.0
Standing with light work	4.0
Walking without load	
on smooth horizontal surface at 2 km hr^{-1}	7.6
on smooth horizontal surface at 3 km hr^{-1}	10.8
on smooth horizontal surface at 4 km hr^{-1}	14.1
on smooth horizontal surface at 5 km hr^{-1}	18.0
on smooth horizontal surface at 6 km hr^{-1}	23.9
on smooth horizontal surface at 7 km hr^{-1}	31.9
on country road at 4 km hr^{-1}	14.2
on grass at 4 km hr^{-1}	14.9
in pine forest, on smooth natural ground at 4 km hr^{-1}	18 to 20
on plowed heavy soil at 4 km hr^{-1}	28.4
Walking and carrying on smooth solid horizontal ground	
1- kg load on back at 4 km hr^{-1}	15.1
30 kg load on back at 4 km hr^{-1}	23.4
50 kg load on back at 4 km hr^{-1}	31.0
100 kg load on back at 3 km hr^{-1}	63.0
Walking downhill on smooth solid ground at 5 km hr^{-1}	
5° decline	8.1
10° decline	9.9
20° decline	13.1
30° decline	17.1
Walking uphill on smooth solid ground at 2.5 km hr^{-1}	
10° incline (gaining height at 7.2 m min^{-1})	
no load	20.6
20 kg on back	25.6
50 kg on back	38.6
16° incline (gaining height at 12 m min^{-1})	
no load	34.9
20 kg on back	44.1
50 kg on back	67.2
25° incline (gaining height at 19.5 m min^{-1})	
no load	55.9
20 kg on back	72.2
50 kg on back	113.8
Climbing stairs 30.5° incline, steps 17.2 cm high,	
100 steps per minute (gaining 17.2 m min^{-1}), no load	57.5
Climbing ladder 70° incline, rungs 17 cm apart	
(gaining 11.2 m min^{-1}), no load	33.6

Note: While Rohmert and Rutenfranz (1983) claim that intra- and inter-individual differences in energy consumption are with ± 10% for the same activity, a comparison of data presented in various texts shows a much higher percentage in variation, particularly at activity levels requiring little energy.

Table 8-5. Total energy cost per day in various jobs and professions (adapted from Astrand and Rodahl 1977). Note that the physical job demands may be different today from what they were decades ago.

Occupation	Energy expenditure, kcal/day		
	Mean	Minimum	Maximum
Men			
Laboratory technicians	2840	2240	3820
Elderly industrial workers	2840	2180	3710
University students	2930	2270	4410
Construction workers	3000	2440	3730
Steel workers	3280	2600	3960
Elderly peasants (Swiss)	3530	2210	5000
Farmers	3550	2450	4670
Coal miners	3660	2970	4560
Forestry workers	3670	2860	4600
Women			
Elderly housewives	1990	1490	2410
Middle-aged housewives	2090	1760	2320
Laboratory technicians	2130	1340	2540
University students	2290	2090	2500
Factory workers	2320	1970	2980
Elderly peasants (Swiss)	2890	2200	3860

Table 8-6. Total energy consumption per kg body weight at various sports activities (adapted from Stagemann 1984).

Sports Activity	Energy Consumption $kJ\ kg^{-1}\ hr^{-1}$
Cross-country skiing, 9 km/hr	38
Jogging, 9 km/hr	40
Ice skating, 21 km/hr	41
Swimming (breaststroke), 3 km/hr	45
Running, 12 km/hr	45
Running, 15 km/hr	51
Wrestling	52
Playing badminton	53
Running, 17 km/hr	60
Bicycling, 43 km/hr	66
Cross-country skiing, 15 km/hr	80

(Classic compilations for many jobs or occupations were done by Durnin and Passmore in 1967 and Spitzer and Hettinger in 1958.) However, when using these tables, one should apply "caution" as explained at the beginning of this chapter.

Instead of using general tables that contain the energy expenditures of, for example, lumbermen, carpenters, housewives, secretaries, etc., one can compose the total energetic cost of given work activities by adding together the energetic cost of the work elements which, combined, make up this activity. This analytic approach largely avoids the problems associated with the fact that the data in the "synthetic tables" (such as Table 8-5) found in older publications reflect measurements taken when many jobs or activities were very different from what they are today: housewives now seldom wash and wring by hand, secretaries do not pound on mechanical typewriters, lumbermen no longer cut down trees with a handsaw and an axe, farmers do not walk behind their plow pulled by oxen or horses, etc. Furthermore, these data may be "averaged" over unknown subjects or attained with only a few subjects and often are inaccurate or unreliable.

If one knows the time spent in a given activity element and its metabolic cost per time unit, one can simply calculate the energy requirements of this element by multiplying its unit metabolic cost with its duration time. To use a simple case: for a person resting (sleeping) eight hours per day, at an energetic cost of approximately 5.1 kJ/min, the total energy cost is about 2450 kJ (5.1 kJ min^{-1} $\times$ 60 min hr^{-1} $\times$ 8 hr). If the person then does six hours of light work while sitting, at 7.4 kJ/min, this adds another 2664 kJ to the energy expenditure. With an additional six hours of light work done standing, at 8.9 kJ/min and further with four hours of walking at 11.0 kJ/min, the total expenditure during the full 24-hr day would come to about 10,960 kJ (or approximately 2610 kcal).

As this example shows, the energetic requirements of given activities, per hour, day, week, or year can be computed from tables of metabolic requirements of certain job elements. In using these, one has to be careful to check whether they include the basal or resting rates or whether they do not. For example, Table 8-4 does not contain the basic value while Tables 8-5 and 8-6 do. In developed countries, daily expenditures range from about 6000 to 20,000 kJ/day, with observed median values of about 10,000 kJ for women and about 14,000 kJ for men.

Energy requirements allow one to judge whether a job is (energetically) easy or hard. Given the largely linear relationship between heart rate and energy uptake, one can often use heart rate to establish the "heaviness" of work. Of course, such labels reflect judgments that rely very much on the current socio-economic concept of what is permissible, acceptable, comfortable, easy, or hard. Depending on the circumstances, one finds a diversity of opinions about how physically demanding a given job is. One such rating of job severity, in terms of energetic or circulatory demands, is presented in Table 8-7. This is a unisex table: most men would find the work lighter, and most women experience the effort heavier than labeled.

"Light" work is associated with rather small energy expenditure (about 10 kJ/min including the basal rate) and accompanied by a heart rate of approximately 90 beats/min. In this type of work, the energy needs of the working muscles are covered by the oxygen available in the blood and by glycogen at the muscle. Lactic acid does not build up. At medium work, with about 20 kJ and 100 beats/min the oxygen requirement at the working muscles is still covered, and initially developed lactic acid is resynthesized to glycogen during the activity. In heavy work, with about 30 kJ and 12 bpm, the oxygen

Table 8-7. Classification of light to heavy work (performed over a whole work shift) according to energy expenditure and heart rate.

| Classification | Total Energy Expenditures | | Heart rate |
	in kJ min^{-1}	in kcal min^{-1}	in beats per minute
light work	10	2.5	90 or less
medium work	20	5	100
heavy work	30	7.5	120
very heavy work	40	10	140
extremely heavy work	50	12.5	160 or more

required is still supplied if the person is physically capable to do such work and specifically trained in this job. However, the lactic acid concentration incurred during the initial minutes of the work is not reduced but remains till the end of the work period, to be brought back to normal levels after cessation of the work.

With light, medium, and even heavy work, metabolic and other physiological functions can attain a steady-state condition throughout the work period (provided the person is capable and trained). This is not the case anymore with very heavy work, where energy expenditures are in the neighborhood of 40 kJ, and heart rate is around 140 beats/min. Here, the original oxygen deficit increases throughout the duration of work, making intermittent rest periods necessary or even forcing the person to stop this work completely. At even higher energy expenditures, such as 50 kJ/min, associated with heart rates of 160 beats/min or higher, lactic acid concentration in the blood and oxygen deficit are of such magnitudes that frequent rest periods are needed and even highly trained and capable persons may be unable to perform this job through a full work shift.

The American Heart Association proposed the MET as the measuring unit for individual work capacity or the metabolic requirement of a task. One MET is defined as the rate of energy expenditure requiring an oxygen consumption of 0.0035 L_{O_2} min^{-1}/kg of body weight, which is close to the basal metabolic rate when sitting (Erb, Fletcher, and Sheffield 1979). This may provide a useful general scale for measuring physical activities; the authors assume that "the average 40-year-old male" in the USA can function at 10 METs at his maximum aerobic capacity.

As discussed earlier, the oxygen supply to the body can rise from about 0.2 L/min^{-1} at rest to 6 L/min^{-1} (in top athletes) with very hard work. This increase to about 30 times the oxygen demand at rest during work is accompanied by a re-direction of the blood supply to the organs in particular need, predominantly to the skeletal muscles which perform physical work. Of the oxygen available at rest, about 20% is usually consumed by muscles, 10% for digestion, 5% by the heart, and around 5% by the brain. At medium work, the muscles will consume about 70% of the oxygen now available, while the digestive tract receives only about 5%. The heart, working harder to supply blood, still receives about 5% of the oxygen available while the brain still consumes about 5%.

Note that only "dynamic work" can be suitably assessed by energy demands. "Static" efforts, where muscles are contracted and kept so, hinder or completely occlude their blood supply (by compression of the capillary bed). Thus, the heart increases its effort to overcome the resistance, showing an increase in heart rate; but because blood flow remains insufficient, relatively little energy is supplied and consumed. Therefore, such static effort is tiring but not well assessed by energy measures. Its fatiguing effect stems from the accrual of metabolites, among them lactic acid and potassium (Kahn and Monod 1989).

Summary of Section 3 "Energy Requirements at Work"

- Energy requirements depend on the activity level of the body

  - to keep the body alive while no effort is done (basal metabolism)
  - to keep the body functioning at rest
  - to maintain the body at work

- At the start of work, both oxygen consumption and heart rate increase sluggishly but heart rate faster, to finally achieve a steady state (at suitable work demand levels). They fall back to resting level, heart rate again faster than oxygen uptake, after cessation of the work.

- If work demands are too high for the individual, no steady rate can be achieved, and the work cannot be continued without interruption for rest.

- Measurement of oxygen consumption (or heart rate) as indicator of the energetic requirements of the work should be performed during the steady-state phase.

- Often, one can calculate the energy requirement of a job by breaking it into elemental subtasks for which the (average) energy requirements are known. In this case, one simply adds the energy needs for each subtask.

- In some cases, energy requirements can be taken from existing listings. However, the given conditions (pertaining e.g., to work details, climate, or worker) may not be reliably the same as assumed for the table values.

- General classifications of job "heaviness" can be done according to ranges of energy and heart rate requirements.

CHAPTER SUMMARY

Calorimetric information can be used for three different purposes:

1) To assess individual energetic functions in response to standardized activities: this is a measurement of individual metabolic work capacity.

2) To assess the energetic requirements of a given activity as reflected by the energetic responses of a given (standard) subject to that work.

3) Combining these two approaches, one can assess the energetic requirements of activities as measured by the energetic responses of a representative sample of workers.

REFERENCES

Astrand, P. 0. and Rodahl, K. 1977 and 1986. *Textbook of Work Physiology*. Second and third editions. New York, NY: McGraw-Hill.

Borg, G. A. V. 1962. *Physical Performance and Perceived Exertion*. Lund: Gleerups.

Borg, G. A. V. 1982. Psychophysical Bases of Perceived Exertion. *Medicine and Science in Sports and Exercise*, 14:5:377–381.

Durnin, J. V. G. A. and Passmore, R. 1967. *Energy, Work, and Leisure*. London: Heinemann.

Eastman Kodak Company. Vol. 1, 1983; Vol. 2, 1986. *Ergonomic Design for People at Work*. New York, NY: Van Nostrand Reinhold.

Ekman, G. 1964. Is the Power Law a Special Case of Fechner's Law? *Perception and Motor Skills*, 19:730.

Erb, B. D., Fletcher, G. F., and Sheffield, T. L. 1979. Standards for Cardiovascular Exercise Treatment Programs. *Circulation*, 59:108A:30– 40.

Fechner, G. T. 1860. *Sachen der Psychophysik*. Leipzig: Breitkopf and Hertel.

Garrow, J. S. 1980. Problems in Measuring Human Energy Balance. In J. M. Kinney (Ed.) *Assessment of Energy Metabolism in Health and Disease* (pp. 2–5). Columbus, OH: Ross Laboratories.

Guyton, A. C. 1979. *Physiology of the Human Body*. Fifth edition. Philadelphia, PA: Saunders.

Kahn, J. F. and Monod, H. 1989. Fatigue Induced by Static Work. *Ergonomics*, 32:7, 839–846.

Kinney, J. M. (Ed.). 1980. *Assessment of Energy Metabolism in Health and Disease*. Columbus, OH: Ross Laboratories.

Louhevaara, V., Ilmarinen, J., and Oja, P. 1985. Comparison of Three Field Methods for Measuring Oxygen Consumption. *Ergonomics*, 28:2:463–470.

Mellerowicz, H. and Smodlaka, V. N. 1981. *Ergometry*. Baltimore: Urban & Schwarzenberg.

Pandolph, K. P. 1983. Advances in the Study and Application of Perceived Exertion. *Exercise and Sport Sciences Review*, 11:118–158.

Pennington, A. J. and Church, H. N. 1985. *Food Values of Portions Commonly Used*. New York: Harper.

Rohmert W. and Rutenfranz J. (Eds.). 1983. *Praktische Arbeitsphysiologie*. Third edition. Stuttgart: Thieme.

Roebuck, J. A., Kroemer, K. H. E., and Thomson, W. G. 1975. *Engineering Anthropometry Methods*. New York, NY: Wiley.

Spitzer, H. and Hettinger, T. 1958. *Tafeln fuer den Kalorienumsatz bei koerperlicher Arbeit*. Darmstadt: REFA.

Stegemann, J. 1981. *Exercise Physiology*. Chicago, IL: Year Book Medical Publications, Thieme.

Stegemann, J. 1984. *Leistungsphysiologie*. Third edition. Stuttgart and New York, NY: Thieme.

Stevens, S. S. 1957. On the Psychophysical Law. *Psychology Review*. 64:151–181.

Weber, E. H. 1834. *De pulse, resorptione, auditu et tactu*. Leipzig, Kochler.

FURTHER READING

Astrand, P. 0. and Rodahl, K. 1986. *Textbook of Work Physiology*. Third edition. New York, NY: McGraw-Hill.

Eastman Kodak Company. Vol. 1 1983; Vol. 2 1986. *Ergonomic Design for People at Work*. New York, NY: Van Nostrand Reinhold.

Pandolph, K. P. 1983. Advances in the Study and Application of Perceived Exertion. *Exercise and Sport Sciences Review*. 11:118–158.

Stegemann, J. 1981. *Exercise Physiology*. Chicago, IL: Year Book Medical Publications, Thieme.

APPENDIX A: TECHNIQUES OF INDIRECT CALORIMETRY

Measuring the oxygen consumed over a sufficiently long period of time is a practical way to assess the metabolic processes. (A physician or physiologist should perform this test.) As discussed earlier, 1 L of oxygen consumed releases about 5 kcal of energy in the metabolic processes. This assumes a normal diet, a healthy body oxidizing primarily carbohydrates and fats under conditions of light to moderate work, and suitable climatic conditions. (The "normality" of the metabolic conditions can be judged, to some degree, by the respiratory exchange quotient, RQ, mentioned earlier.)

Classically, indirect calorimetry has been performed by collecting all exhaled air during the observation period in airtight bags (Douglas Bags). The volume of the exhaled air is then measured and analyzed for oxygen and carbon dioxide as needed for the determination of the RQ. From these data one can calculate the amount of energy used during the collection period. This requires a rather complex air collecting system, including nose clip and intake and exhaust valves which mostly limits this procedure to the laboratory. A major improvement was done by diverting only a known percentage of the exhaled air into a small collection bag. This procedure (developed in the Max-Planck Institute of Work Physiology by Mueller and Franz) is still in use since only a relatively small device has to be carried by the subject, thus allowing performance of most daily activities without much hindrance by the collecting system. Still, in both cases the subject must wear a face mask which serves to measure the total amount of air ventilated, and which separates the exhaled from the inhaled or ambient air. This can become quite uncomfortable for the subject and hinders speaking.

Significant advances have been made in the last decade through the use of instantaneously reacting oxygen sensors which can be placed into the air flow of the exhaled air, allowing a breath-by-breath analysis. The differences in oxygen measured with different equipment are usually rather small; for example, a comparison between the classical Douglas Bag method, the Max-Planck gas meter and the Oxylog®, a small portable instrument, showed variations in the mean of less than 7%; the linear regression coefficients were better than 0.90 (Louhevaara, Ilmarinen, and Oja 1985). For most field observations, the accuracy of the bagless procedures are quite sufficient. Since the volume of exhaled air can also be measured by suitable sensors, more recently "open" face masks have been developed that draw a stream of air across the face of the subject who simply inhales from this air flow and exhales into it. This allows free breathing, free speech, and even cools the face (which might in fact influence the working capacity to a small extent).

APPENDIX B: RATING THE PERCEIVED EFFORT

Around the middle of the 19th century, models of the relationships between a physical stimulus and one's perceptual sensation of that stimulus, i.e., the "psychophysical correlate" were developed. Weber (1834) suggested that the "just noticeable difference ΔI" which can be perceived depends on the absolute magnitude of the physical stimulus I:

$$\Delta I = \alpha\ I \qquad (8\text{-}7)$$

where α is a constant.

Fechner (1860) related the magnitude of the "perceived sensation P" to the magnitude of the stimulus:

$$P = \beta + \gamma\ \log I \qquad (8\text{-}8)$$

where β and γ are constants.

In the 1950s, S. S. Stevens at Harvard and G. Ekman in Sweden introduced "ratio scales" (assuming a zero point and equidistant scale values), which since have been used to describe the relationships between the perceived intensity and the physically measured intensity of a stimulus in a variety of sensory modalities (e.g. related to sound, lighting, climate) as:

$$P = \delta\ I^n \qquad (8\text{-}9)$$

where δ is a constant and n ranges from 0.5 to 4, depending on the modality.

Since about 1960, Borg and his collaborators have modified these relationships to take into account deviations from previous assumptions (such as zero point and equidistance), and to describe the perception of different kinds of physical efforts. Borg's "General Function" is:

$$P = a + c\,(I - b)^n \qquad (8\text{-}10)$$

with the constant 'a' representing "the basic conceptual noise" (normally below 10% of I) and the constant 'b' indicating the starting point of the curve; 'c' is a conversion factor dependent on the type of effort.

Ratio scales indicate only proportions between percepts, but do not indicate absolute intensity levels; they neither allow intermodal comparisons nor comparisons between intensities perceived by different individuals. Borg has tried to overcome this problem by assuming that the subjective range and intensity level are about the same for each subject at the level of maximum intensity. In 1960, this lead to the development of a "category scale" for the Rating of Perceived Exertions (RPE). The scale ranges from 6 to 20 (to match heart rates from 60 to 200 beats/min). Every second number is anchored by verbal expressions:

The 1960 Borg RPE Scale (modified 1985)

6	–	(no exertion at all)
7	–	extremely light
8		
9	–	very light
10		
11	–	light
12		
13	–	somewhat hard
14		
15	–	hard (heavy)
16		
17	–	very hard
18		
19	–	extremely hard
20	–	maximal exertion

In 1980, Borg proposed a "category scale with ratio properties" which could yield ratios, levels, and allow comparisons but still retain the same correlation (of about 0.88) with heart rate as the RPE scale, particularly if large muscles were involved in the effort.

The Borg General Scale (1980)

0	–	nothing at all
0.5	–	extremely weak (just noticeable)
1	–	very weak
2	–	weak
3	–	moderate
4	–	somewhat strong
5	–	strong
6	–	
7	–	very strong
8	–	
9	–	
10	–	extremely strong (almost maximal)

(*Note*: The terms "weak" and "strong" may be replaced by "light," and "hard," or "heavy," respectively).

The instructions for use of the scale as are follows (modified from Borg's publications):

While the subject looks at the rating scale:

"I will not ask you to specify the feeling, but do select a number which most accurately corresponds to your perception of (experimenter specifies symptom)

If you don't feel anything, for example, if there is no (symptom), you answer *zero* — nothing at all.

If you start feeling something, just about noticeable, you answer *0.5* — extremely weak, just noticeable.

If you have an extremely strong feeling of (symptom) you answer *10* — extremely strong, almost maximal. This is the absolute strongest which you have every experienced.

The more you feel, the stronger the feeling, the higher the number which you choose.

Keep in mind that there are no wrong numbers; be honest; do not overestimate or underestimate your ratings. Do not think of any other sensation than the one I ask you about.

Do you have any questions?"

Let the subject get well acquainted with the rating scale before the test. During the test, let the subject do the ratings towards the end of every work period, i.e. about 30 seconds before stopping or changing the workload. If the test must be stopped before the scheduled end of the work period, let the subject rate the feeling at the moment of stoppage.

Chapter 9

INTERACTIONS OF THE BODY
WITH THE THERMAL ENVIRONMENT

OVERVIEW

This chapter discusses the heat generation within the body in relation to the energy emitted from it via external work done and by evaporation, and to the heat exchanged (received or dissipated) by radiation, convection, and conduction. The physiologic and physical means to effect this energy transfer, and the engineering control of the macro- and micro-climate are explained. Recommendations for suitable ergonomic conditions are given.

The Model

The human body generates energy and exchanges (gains or loses) energy with the environment. Since a rather constant core temperature must be maintained, suitable heat flow from the environment to the body, or from the body outwards, must be achieved. The internal energy flow is primarily controlled in the body masses between skin and core.

THE HUMAN BODY AS A THERMO-REGULATED SYSTEM

The human body has a complex control system to maintain the body core temperature near 37°C (about 99°F) with a variation of just a few degrees. While there is some fluctuation throughout the day due to diurnal changes in body functions, the main impact upon the human thermal regulatory system results from the interaction between (metabolic) heat generated within the body and external energy gained in hot surroundings or lost in a cool environment. If the deep body temperature deviates just a few degrees from its set value, physical and mental work capacities are impaired: cellular structures, enzyme systems, and many other functions are directly affected by changes in body temperature.

If the temperature at a human cell exceeds 45°C, heat coagulation of proteins takes place, but if the temperature reaches freezing, ice crystals break the cell apart. In its effort to protect itself from conditions that are either too hot or too cold, the human temperature regulation system must keep temperatures well above freezing and below the 40s in its outer layers. At the core, a range close to 37°C must be maintained; changes in core temperature of ±2°C from 37°C affect body functions and task performance severely, while deviations of ±6°C are usually lethal (ASHRAE 1985).

The Energy Balance

In Chapter 8 on the metabolic system, the energy equation between (nutritional) inputs and outputs was given as

$$I = W + M + S \tag{9-1}$$

where I is the energy input via nutrition, W the external work done, M the metabolic heat generated, and S the energy storage in the body.

Assuming for convenience that the quantities I and W remain unchanged, one can concentrate on the energy exchange with the thermal environment

$$I - W = \text{constant} = M + S \tag{9-2}$$

The system is in balance with the environment if all metabolic energy M is dissipated to the environment without a change in the quantity S. If heat storage S increases, not all metabolic energy could be dispelled to the environs and/or energy was transferred from the environment to the body. If S becomes smaller, more than M was lost from the body to the environment.

ENERGY EXCHANGES WITH THE ENVIRONMENT

Energy is exchanged with the environment through *radiation* R, *convection* C, *conduction* K, and *evaporation* E.

Heat exchange through *radiation* R depends primarily on the temperature difference between two opposing surfaces, for example between a window pane and a person's skin. Heat is always radiated from the warmer to the colder surface; for example to the cold window in the winter or to the body from a sun-heated pane in the summer. Therefore, the body can either lose or gain heat through radiation. This radiative heat exchange does not depend on the temperature of the air between the two opposing surfaces.

The amount of radiating energy Q (per second) gained (+) or lost (−) by the human body through radiation is (according to Stegemann 1984)

$$Q_R = a\ s\ (\delta T_o^4 - \epsilon T^4)\ \text{in J s}^{-1} \tag{9-3}$$

with a = 75 Js^{-1}m^{-2} deg K^{-4} (Radiation Constant)
     s = body surface participating in the energy exchange, in m^2
     δ = absorption coefficient (see below)
     T_o = temperature of opposing surface, in deg K
     ϵ = emission coefficient (see below)
     T = body surface temperature, in deg K.

The wave lengths of radiation from the human body are $3 < \lambda < 60\ \mu m$, i.e., in the infrared range. Hence, it radiates like a black body, i.e., with an emission coefficient ϵ close to 1, independent of the color of the radiating human skin. However, the absorption coefficient δ does depend on skin color; for solar rays (with wavelengths $0.3 < \lambda < 4\ \mu m$) it ranges from 0.6 for light skinned people to 0.8 for dark skinned persons.

Energy is also exchanged through *convection* C and *conduction* K. In each case, the heat transferred is proportional to the area of human skin participating in the process, and to the temperature difference between skin and the adjacent layer of the external medium. In general terms, heat exchange per second (by convection or conductance) is

$$Q_{C,K} = f\ [s\ h\ (t_m - t)] \tag{9-4}$$

with h = heat conduction coefficient (see below)
     s = body surface participating in the heat exchange, in m^2
     t_m = temperature of the medium with which s is in contact, in °C
     t = temperature of the body surface s, in °C.

The heat conduction coefficient h of human tissue is $3 < h < 260$ J cm^{-1} s^{-1} °C^{-1}, i.e., the amount of energy that penetrates tissue 1-cm thick per second, when the temperature difference is 1°C.

Conductance K exists when the skin contacts a solid body, such as a piece of iron. Energy flows from the warmer body to the colder one; as the temperatures of the contact surface become equal, the energy exchange ceases. The rate and amount of heat exchanged also depend on the conductance of the touching bodies. Cork or wood "feel warm" because their heat conduction coefficients are below that of human tissue, but cool metal accepts body heat easily and conducts it away.

Exchange of heat through *convection* C takes place when the human skin is in contact with air and fluids, e.g. water. Like in conduction, heat energy from the skin is transferred to a colder gas or fluid next to the skin surface or transferred to the skin if the surrounding medium is warmer. Convective heat exchange is facilitated if the medium moves quickly along the skin surface (in laminar or more often turbulent fashion), thus maintaining a temperature differential.

The method of exchanging heat via convection Q_C, gain (+) loss (−), is similar to conduction; but the effect of moving the immediate layer of the surrounding medium modifies the process

$$Q_C = c\ s\ (t_m - t) \tag{9-5}$$

with the convection coefficient in general $21 < c < 37$ kJ hr^{-1} m^{-2} °C^{-1} (Stegemann 1984). It is much dependent on the actual relative movement of the medium. As long as there is a temperature gradient between the skin and the medium, there is always some natural movement of air or fluid: this is called "free convection." Much more movement can be produced by forced action (e.g. by an air fan, or while swimming in water rather than floating motionless): this is called "induced convection."

For the nude body in water, the heat loss is (Nadel, 1984)

$$Q_c = h_c \, s \, (t_m - t) \quad \text{for } t_m < t \tag{9-6}$$

with the convective transfer coefficient h_c at about 230 $\text{Wm}^{-2}\text{°C}^{-1}$ during rest in still water; it is about 580 $\text{Wm}^{-2}\text{°C}^{-1}$ when swimming at any speed because of the turbulence of the water layer near the body produced by the swimming activities.

Heat exchange by *evaporation* is only in one direction. The human loses heat by evaporation; there is no condensation of water on the skin, which would add heat. Evaporation of water (sweat) on the skin requires an energy of about 2440 J (580 cal) per cm^3 of evaporated fluid, which reduces the heat content of the body by that amount.

The heat lost by evaporation Q_E from the human body is a function of participating wet body surface, humidity and vapor pressures (according to Nadel and Horvath 1975)

$$Q_E = f \, [s \, (h_r p_a - P)] \tag{9-7}$$

with s = body surface participating in the heat dispersion
 h_r = relative humidity of the surrounding air
 p_a = vapor pressure in the surrounding air
 p = vapor pressure at the skin.

Of course, Q_E is zero for $p \leq p_a$ since there can only be heat loss by evaporation if the humidity of the surrounding air is less at the skin. Therefore, movement of the air layer at the skin (convection) increases the actual heat loss through evaporation if this replaces humid air by dryer air.

Some evaporative heat loss occurs even in a cold environment because there is always evaporation of water in the lungs that increases with enlarged ventilation at work, and there is also secretion of some sweat onto the skin surface in physical work. The nude body at rest in the cold loses 6 to 10 Wm^{-2} from the skin and 3 to 6 Wm^{-2} from the respiratory tract (Nadel and Horvath 1975).

Given these variables, heat balance exists when metabolic energy M developed in the body, heat storage S in the body, and heat exchanges with the environment by radiation R, convection C, conduction K, and evaporation E are in equilibrium. This can be expressed as

$$M + S + R + C + K + E = O \tag{9-8}$$

The quantities R, C, K, and E are negative if the body loses energy to the environment (E can only be negative) and positive if the body gains energy from the environment.

TEMPERATURE REGULATION AND SENSATION

Heat energy is circulated throughout the body by the blood. The blood flow is modulated by the vasomotor actions of constriction, dilation, and shunting. Heat is exchanged with the environment at the body's respiratory surfaces, i.e., in the lungs and at respiratory mucosa and of course through the skin.

Heat is produced in the body's "metabolically active" tissues: primarily at skeletal muscles, but also in internal organs, fat, bone, connective and nerve tissue. In a cold environment, heat must be conserved, which is primarily done by the reduction of blood flow to the skin and by increased insulation. In a hot environment body heat must be dissipated and gain from the environment be prevented. This is primarily done by increased blood flow to the skin, by sweat production and by evaporation.

The body must regulate its temperature to prevent undercooling or overheating. The temperature of key tissues, such as brain, heart, lungs, and abdominal organs, must be kept rather constant. However, the body temperature is not at all uniform; there are large temperature differences between the "core" and the "shell." Under normal conditions, the average gradient between skin and deep body is about 4°C at rest, but in the cold the difference in temperature may be 20°C or more. Even within the core the temperature varies by several degrees. This makes it rather difficult to speak about "one" body temperature since it is so variable throughout the body. In reality, the temperature regulation system has to maintain various temperatures at various locations under different conditions.

Figure 9-1 shows a model of the regulation of the human energy balance. It breaks the actual regulatory system into three subsystems: the controlling, the effecting, and the regulated elements. The human body has given set points, near 37°C in the brain and about 33°C at the skin. Any deviations from these values are detected by various temperature sensors, and counteractions are initiated at the hypothalamus. Here, preoptic neurons affect three different pathways: the efferent nervous system to change the muscle tonus, the sudomotor system to affect sweat production, and the vasomotor system to bring about vascular dilation, constriction, or shunting. The effecting system interacts with the work or exercise being performed. For example, if less heat must be generated internally, muscular activities will be reduced, possibly to the extent that no

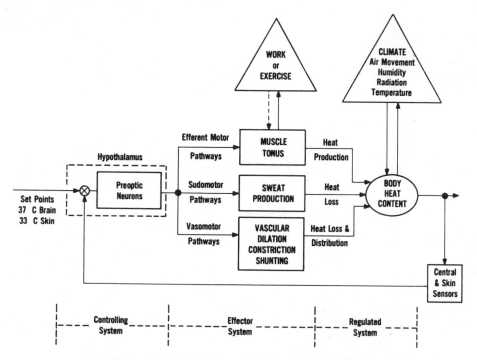

Figure 9-1. Model of the regulation of body heat content.

work is being performed anymore. On the other hand, if more heat must be generated the work or exercise level will be augmented by increased muscular activities. (Given the low efficiency of muscular work, it generates much heat.) However, muscle activities can generate only more or less heat but cannot cool the body. In contrast, sweat production only influences the amount of energy lost but cannot bring about a heat gain. Vascular activities can affect the heat distribution through the body and control heat loss or gain, but they do not generate energy.

Muscular, vascular, and sweat production functions regulate the body heat content in direct interaction with the external climate. The climate itself is defined by humidity, radiation, temperature, and air (or fluid) movement, as discussed later.

Various temperature sensors are located in the core and the shell of the body. Hot sensors generate signals (sent to the hypothalamus), particularly in the range of approximately 38 to 43°C. On the other hand, the major sensitivity to cold is from about 15 to 35°C. There is some overlap in the sensations of "cool" and "warm" in the intermediate range. Between about 15 and 45°C, our perception of either "cold" or "hot" condition is highly adaptable. Below 15°C and above 45°C the human temperature sensors are less discriminating but also less adapting. A "paradoxical" effect is that, around 45°C, sensors again signal "cold" while in fact the temperature is rather hot.

Deviations in the actual temperatures from the set points (in the brain and skin) bring about counteractions by the controlling system. However, the set points are variable by several degrees and change, for example, throughout the course of the day with diurnal rhythms, or with acclimatization.

THERMAL HOMEOSTASIS

The human regulatory system must achieve two suitable temperature gradients: from the *core to the skin* and from the *skin to the surroundings*. The gradient from the core to the skin is the most important because overheating or undercooling of the key tissues in the brain and the trunk must be avoided, even at the cost of overheating or undercooling the shell. In the central body parts this gradient should not exceed approximately 4°C. The primary means to effect this important core/skin gradient are located in the more peripheral tissues, i.e., in the skeletal muscles and in the superficial arteries and veins and their connecting systems.

Thermal homeostasis is *primarily* achieved by regulation of the blood flow from deep tissues and muscles to lungs and skin.

The ability to absorb and transport heat is 3 to 3.5 $Jg^{-1}°C$ in tissue but reaches 4 $Jg^{-1}°C^{-1}$ in the blood. Blood is a relatively efficient way to transport heat within the body. In the lungs, 10% to 25% of the total dissipated heat is transmitted to the environment; most heat is exchanged at the skin.

Secondary activities to establish thermal homeostasis take place at the muscles, either by voluntary activities (changes in external work and exercise) or by involuntary shivering. Here, different actions are taken depending on the goal of the regulatory system: If heat gain is to be achieved, skeletal muscle contractions are initiated; but if heat loss is desired, muscular activities are abolished. (Also, loss is achieved through regulation of skin blood supply, sweat production, and changes in clothing and shelter.)

If a heat gain must be prevented, clothing and shelter are adjusted; if heat loss must be prevented, vasoconstriction methods at the skin will be used, and clothing and shelter will be altered.

Changes in clothing and shelter are *tertiary* actions to achieve thermal homeostasis. They affect radiation, convection, conduction, and evaporation. "Light" or "heavy" clothes have different permeability and ability to establish stationary insulating layers. Clothes affect conductance, i.e., energy transmitted per surface unit, time, and temperature gradient. Also, their color determines how much external radiation energy is absorbed or reflected. Similar effects are brought about by shelters, which by their material, distance from the body, form and color determine whether heat is gained or lost by the body through radiation, convection, and evaporation.

MEASUREMENT OF BODY TEMPERATURES

One way of assessing the exchange of heat with the environment is to perform direct calorimetry, where a person is placed into an energy-tight compartment which allows the measurement of all heat energies exchanged — see Chapter 8. However, this is a tedious procedure which severely limits the ability of the person to work "normally." Most methods to assess the heat balance are based on temperature measurements within or at the body.

The so-called safe temperature range of the core is between 35°C to 40°C. Changes in core temperatures, particularly severe deviations from the "safe" temperatures indicate overloads or dysfunctions of the energy regulatory mechanisms in the body.

A variety of techniques exist to measure the temperature in the body parts. For example, inserting a temperature probe into the esophagus or the stomach allows measurement of deep body temperatures. Another "classical" site is the rectum, where temperature probes are usually inserted 5 to 10 cm behind the sphincter. With the hypothalamus providing the reference temperature, the rectal temperature is usually about 0.5°C lower, provided that a steady state has been maintained for about 30 minutes. The rectal temperature of a resting individual is slightly higher than the temperature of arterial blood and about the same as in the liver. Brain temperature rises more quickly in response to heat influx than rectal temperature. Temperatures measured at the ear drum follow actual brain temperatures rather closely as do the temperatures measured in the esophagus. Temperature measurement in the mouth or in the armpit is less accurate but rather easily done and socially acceptable.

As discussed earlier, skin temperatures may vary much from core temperatures; they can be 20°C below or 10°C above the temperatures at the hypothalamus. Measurement of temperatures at various skin sites therefore, provides information that is loosely related to core temperatures. Furthermore, various sites on the body surface may be at very different temperatures. Hence, the concept of "average skin temperature" is a difficult one; however it has been used by assigning weighting factors to the measurements taken at various body surfaces, depending on the proportions of these surface areas compared to the total body. For example, the "Hardy and Dubois procedure" is to multiply measurements taken at the head by 0.07, the arms by 0.14, the hands by 0.05, the trunk by 0.35, the thighs by 0.19, the lower legs by 0.13, and at the feet by 0.07. The results are added for an "average" skin temperature.

Various procedures exist to measure and calculate the mechanisms of heat regulation. For example, if one establishes the "average" skin temperature $\overline{T}_s$, measures the rectal temperature T_r and knows the body mass W in kg, one can calculate the heat content from the "Burton equation"

$$\text{Heat content} = 3.47 \ W \ (0.65 \ T_r + 0.35 \ \overline{T}_s) \tag{9-9}$$

However, the specific heat of the body, assumed to be 3.47 kJ kg^{-1} °C^{-1} may vary from 2.93 to 3.45, depending on the individual's body composition. Also, the ratios of T_r and

T_s assumed to be 0.65 and 0.35, respectively, are not fixed but range from 0.9 and 0.1 in warm environments and during exercise to 0.6 and 0.4 at rest in a cool environment (Stolwijk 1980).

ASSESSMENT OF THE THERMAL ENVIRONMENT

The thermal environment is determined by four physical factors: air (or water) temperature, humidity, air (or water) movement, and temperature of surfaces. The combination of these four factors determines the physical conditions of the climate and our perception of the climate.

Measurement of the *air temperature* is performed with thermometers, usually filled with alcohol or mercury. Of course, other techniques using thermistors or thermocouples can be used as well. However, whichever technique is used, it must be ensured that the ambient temperature is not affected by the other three climate factors (humidity, air movement, and surface temperatures). To measure the so-called "dry temperature" of ambient air, one keeps the sensor dry and shields it with a surrounding bulb that reflects radiated energy. (Hence, air temperature is often measured with a so-called "dry bulb" thermometer.)

The *air humidity* may be measured with a psychrometer, hygrometer, or other electronic devices. These usually rely on the fact that the cooling effect of evaporation is directly proportional to the humidity of the air, with higher vapor pressure making evaporative cooling less efficient. Air humidity may be expressed either in absolute or in relative terms. The highest absolute content of vapor in the air is reached when any further increase would lead to the development of water droplets, falling out of the gas. The amount of possible air vapor depends on the temperature of the air, with higher temperatures allowing more water vapor to be retained than at lower temperatures. One usually speaks of "relative humidity," which indicates the actual vapor content in relation to the possible maximal content at the given air temperature and air pressure.

Air movement is measured with various types of anemometers, usually based on mechanical or electrical principles. One may also measure air movement with two thermometers — one dry and one wet (similar to what can be done to assess humidity), relying on the fact that the wet thermometer shows more increased evaporative cooling with higher air movement than the dry thermometer. Air movement helps particularly in convective heat exchange, because it moves "fresh" air to skin surfaces. Here, turbulent air movement is as effective as laminar movement in heat transfer.

Radiant heat exchange depends primarily on the difference in temperatures between the individual and the surroundings, on the emission properties of the radiating surface, and on the absorption characteristics of the receiving surface. While there is no problem in measuring surface temperatures, one easy way to assess the amount of energy transferred through radiation is to place the thermometer inside a black globe, which absorbs practically all radiated energy.

Various techniques exist to express the combined effects of the four environmental factors in one model, chart, or index. For example, the outdoors WBGT (Wet Bulb Globe Temperature) weighs the effects of several climate parameters

$$WBGT = 0.7\, NWBT + 0.2\, GT + 0.1\, DBT \qquad (9\text{-}10)$$

where

NWBT is the Natural Wet Bulb Temperature of a sensor in a wet wick exposed to
   natural air current;

GT is the Globe Temperature at the center of a black sphere of 15 cm diameter; and

DBT is the Dry Bulb Temperature measured while shielded from radiation.

Various devices are on the market which combine several measurements into one
number, including the "effective temperature" discussed later.

REACTIONS OF THE BODY TO COLD ENVIRONMENTS

In a cold climate, the body must conserve heat while producing it. For this, there are
two major ways to regulate the temperature: distribution of the blood flow, and increase
in metabolic rates.

To conserve heat, the temperature of the skin is lowered to reduce the temperature
difference against the outside. This is done by displacing the circulating blood towards
the core, away from the skin. This can be rather dramatic; for example, the blood flow in
the fingers may be reduced to 1% of what existed in a moderate climate.

Blood distribution can be regulated by three procedures: constriction of skin vessels,
use of deep veins, and increased heat exchange between arteries and veins.

In a resting individual, lightly clad, in an "ideal" external temperature of about 28°C,
the mean skin temperature is about 33°C and the core temperature about 37°C. This
temperature gradient, from core to skin, allows the transfer of excess heat from the
metabolically active tissues to the environment. Of the total circulating blood volume
(about 5 L) about 5% flows through the blood vessels in the skin. A cold environment
constitutes a large temperature difference to the skin, which would cause increased heat
loss through convection and radiation. By closing much of the blood vessel pathways,
less blood flows towards the superficial skin surfaces; the skin temperature is lowered
and hence the energy flow towards the environment is reduced. One may consider this as
a reduction in the "conductance of the surface tissues." The vasoconstriction of the skin
blood vessels is usually accompanied by the second method to reduce tissue conductance:
blood in the veins of the extremities and near the skin, is re-routed from the superficial to
the deep veins. These deep veins are anatomically close to the arteries, which carry warm
blood from the heart. Therefore, a heat exchange between the arteries and veins in the
deep body tissues occurs. Having cooled arterial blood supplying the skin or extremities
brings about two effects: cooling of the body core is reduced, and the extremities and
surfaces are cool, thus having less conductance of heat to the outside.

This displacement of the blood volume from the skin to the central circulation is very
efficient in keeping the core warm and the surfaces cold. Peripheral vasoconstriction can
bring about a six-fold increase in the insulating capacity of the subcutaneous tissues,
accompanied by the earlier mentioned reduction of the blood volume to as little as 1%.
The danger associated with these vasoconstrictional regulatory actions is that the
temperature in the peripheral tissues may approach that of the environment. Thus, cold
fingers and toes may result, with possible damage to the tissue if the temperatures get
close to freezing. The blood vessels of the head do not undergo as much vasoconstriction
so the head stays warm even in cold environments, with less danger to the tissues;
however, the resulting large difference in temperature to the environment brings about a
large heat loss, which can be prevented by wearing a hat or scarf to create an insulating
layer.

Incidentally, the development of "goose bumps" of the skin helps to retain a layer of stationary air close to the skin, which is relatively warm and has the effect of an insulating envelope, reducing energy loss at the skin.

The mechanism of blood redistribution in the cold indicates, again, the overriding intent to keep the core temperature high enough, even at the risk of cooling the shell to the extent that local damage may occur there.

The other major reaction of the body to a cold environment is the increase of metabolic heat generation. This may occur involuntarily, i.e., by reflex stimulation of rapid movements of antagonistic muscle groups against each other, known as shivering. Since no mechanical work is done to the outside, the total activity is transformed into heat production, allowing an increase in the metabolic rate to up to 4 times the resting rate. Of course, muscular activities also can be done voluntarily, such as by either increasing the dynamic muscular work performed, or by moving body segments, contracting muscles, flexing the fingers, etc. Such dynamic muscular work may easily increase the metabolic rate to ten times or more than that at rest.

Indices of Cold Strain

If vasoconstriction and metabolic rate regulation cannot prevent serious energy loss through the body surfaces, the body will suffer some effects of cold stress. The skin is, as just discussed, first subjected to cold damage while the body core is protected as long as possible.

As the skin temperature is lowered to about 15°C to 20°C, manual dexterity begins to be reduced. Tactile sensitivity is severely diminished as the skin temperature falls below 8°C. If the temperature approaches freezing, ice crystals develop in the cells and destroys them, a result known as "frost bite." Reduction of core temperature is more serious, where vigilance may begin to drop at temperatures below 36°C. At core temperatures of 35°C, one may not be able to perform even simple activities. When the core temperature drops even lower, the mind becomes confused, with loss of consciousness occurring around 32°C. At core temperatures of about 26°C, heart failure may occur. At very low core temperatures, such as 20°C, vital signs disappear, but the oxygen supply to the brain may still be sufficient to allow revival of the body from hypothermia.

Of course, severe reductions in skin temperatures are accompanied by a fall in core temperature. At local temperatures of 8°C to 10°C, peripheral motor nerve velocity is decreased to near zero; this generates a nervous block, which helps to understand why local cooling is accompanied by rapid onset of physical impairment. Severe cooling of the skin and central body goes along with increasing inability to perform activities, even if they could save the person ("cannot light a match") leading to apathy ("let me sleep") and final hypothermia.

Hypothermia can occur very quickly if a person is exposed to cold water. While one can endure up to 2 hours in water at 15°C, one is helpless in water of 5°C after 20 to 30 minutes. The survival time in cold water can be increased by wearing clothing which provides insulation; also, obese persons with much insulating adipose tissue are at advantage over skinny ones. Floating motionless results in less metabolic energy generated and spent than when swimming vigorously.

REACTIONS OF THE BODY TO HOT ENVIRONMENTS

In hot environments, the body produces heat and must dissipate it. As in cold environments, two primary means exist to control the energy flow: blood distribution and metabolic rate. Now, however, the body must dissipate heat instead of preventing heat loss. To achieve this, the skin temperature should be near, best above the immediate environment.

Blood is redistributed to allow heat transfer to the skin. For this, the skin vessels are dilated and the superficial veins fully opened, actions directly contrary to the ones taken in the cold. This may bring about a fourfold increase in blood flow above the resting level, increasing the conductance of the tissue. Accordingly, energy loss through convection, conduction, and radiation (which all follow the temperature differential between skin and environment) is facilitated.

If heat transfer is still not sufficient, sweat glands are activated, and the evaporation of the produced sweat cools the skin. Recruitment of sweat glands from different areas of the body varies among individuals. Some persons have few sweat glands, while most have at least 2 million sweat glands in the skin so that large differences in the ability to sweat exist among individuals. The activity of each sweat gland is cyclic. The overall amount of sweat developed and evaporated depends very much on clothing, environment, work requirements, and on the individual's acclimatization.

If heat transfer by blood distribution and sweat evaporation is insufficient, muscular activities must be reduced to lower the amount of energy generated through metabolic processes. In fact, this is the final and necessary action of the body if otherwise the core temperature would exceed a tolerable limit. If the body has to choose between unacceptable overheating and continuing to perform physical work, the choice will be in favor of core temperature maintenance, which means reduction or cessation of work activities.

Indices of Heat Strain

There are several signs of excessive heat strain on the body. The first one is the sweat rate. Above the so-called insensible perspiration (in the neighborhood of about 50 cm^3 hr^{-1}) sweat production increases depending on the heat that must be dissipated. In strenuous exercises and hot climates, several liters of sweat may be produced in one hour. However, on the average during the working time usually not more than about 1 L/hr is produced, but sweat losses up to 12 L in 24 hrs have been reported under extreme conditions. Sweat begins to drip off the skin when the sweat generation has reached about 1/3 of the maximal evaporative capacity. Of course, sweat running down the skin contributes very little to heat transfer.

Increases in the circulatory activities signal heat strain. Cardiac output must be enlarged, which is mostly brought about by a higher heart rate. This may be associated with a reduction in systolic blood pressure. Another sign of heat strain is a rise in core temperature, which must be counteracted before the temperature exceeds the sustainable limit.

The water balance within the body provides another sign of heat strain. Dehydration indicated by the loss of only one or two percent of body weight can critically affect the ability of the body to control its functions. Therefore, the fluid level must be maintained — by frequent drinking of small amounts of water. Sweat contains different salts, particularly NaCl, in smaller concentrations than in the blood. Sweating, which extracts water from the plasma, augments the relative salt content of the blood. Normally, it is not necessary to add salt to drinking water since in western diets the salt in the food is more than sufficient to resupply the salt lost with the sweat.

Water supply to the body comes from fluids drunk, water contained in food, and water chemically liberated during oxidation of nutrients. Daily water losses are approximately: from the gastrointestinal tract, 0.2 L; from the respiratory tract, 0.4 L; through the skin,

0.5 L; from the kidneys, 1.5 L. Obviously, these figures can change considerably when a person performs work in a hot environment.

Among the first reactions to heavy exercise in excessive heat are sensations of discomfort and perhaps skin eruptions (*prickly heat*) associated with sweating. As a result of sweating, so-called *heat cramps* may develop, which are muscle spasms related to local lack of salt. They may occur after quickly drinking large amounts of fluid.

So-called *heat exhaustion* is a combined function of dehydration and overloading the circulatory system. Associated effects are fatigue, headache, nausea, dizziness, often accompanied by giddy behavior. *Heat syncope* indicates a failure of the circulatory system, demonstrated by fainting. *Heat stroke* indicates an overloading of both the circulatory and sweating systems and is associated with hot dry skin, increased core temperature, and mental confusion. Table 9-1 lists symptoms, causes, and treatment of heat stress disorders.

SUMMARY OF THE THERMOREGULATORY ACTIONS

While working in a hot or cold environment, the primary purpose of the human thermoregulatory system is to keep the body core temperature (energy content) within narrow limits. For this, a suitable temperature differential (energy flow) must be established between the deep body tissues and the skin. To maintain the core temperature, the body uses a number of procedures:

To *achieve* heat *gain*, such as in a cold environment, metabolism is increased through contractions of skeletal muscle.

To *achieve* heat *loss*, blood supply to the skin and sweat production are increased.

To *prevent* heat *gain*, muscular activities and metabolic functions are reduced.

To *prevent* heat *loss*, blood supply to the skin is reduced and muscle contractions are increased.

VARIABILITY OF HEAT GENERATION AND HEAT DISTRIBUTION IN THE BODY

All metabolic functions of the body at rest finally result in the generation of heat. *Assuming no interaction* with the environment for the moment, the basal metabolism generates approximately 100 J/sec. This heat generation may multiply with increased activity. For example, a highly trained 70-kg bicyclist may have an oxygen capacity of 5 L/min. With an energy efficiency of 20%, 80% of the oxygen consumed serves to develop metabolic heat; in this case, approximately 1.4 kJ/sec^{-1} are generated. With no heat exchange, the body temperature would increase quickly. The average heat capacity of the human body is approximately 3.4 kJ kg^{-1}/°C. (This means that 3.4 kJ are needed to increase the temperature of 1 kg of tissue by 1°C.) The exemplary bicyclists would increase his total body temperature nearly 1.5°C in 1 hr if resting, while this same temperature increase would be achieved within about 3 min when bicyling hard, if there were no heat exchange with the environs.

Of course, this example is very much simplified. In reality, only the core temperature is kept constant while the temperatures of the body shell can be varied. Assuming that the shell mass is about 1/3 of the total body mass and that its temperature may be increased by 6°C without ill effects, the total body energy could be increased by approximately 500 kJ without affecting the core temperature.

Table 9-1. Heat stress disorders (adapted from Spain, Ewing, and Clay 1985).

	SYMPTOMS	CAUSES	TREATMENTS
Transient Heat Fatigue	Decrease in productivity, alertness, coordination and vigilance.	Not acclimatized to hot environment.	Graduate adjustment to hot environment.
Heat Rash ("Prickly Heat")	Rash in area of heavy perspiration; discomfort; or temporary disability.	Perspiration not removed from skin; sweat glands inflamed.	Periodic rests in a cool area; showering/bathing; drying skin.
Fainting	Blackout, collapse.	Shortage of oxygen in the brain.	Lay down.
Heat Cramps	Painful spasms of used skeletal muscles.	Loss of salt; large quantities of water consumed quickly.	Adequate salt with meals; salted liquids (unless advised differently by a physician).
Heat Exhaustion	Extreme weakness or fatigue; giddiness; nausea; headache; pale or flushed complexion; body temperature normal or slightly higher; moist skin; in extreme cases vomiting and/or loss of consciousness.	Loss of water and/or salt; loss of blood plasma; strain on the circulatory system.	Rest in cool area; salted liquids (unless advised differently by a physician).
Heat Stroke	Skin is hot, dry and often red or spotted; core temperature is 40°C (105F) or higher and rising; mental confusion; deliriousness; convulsions; possible unconsciousness. Death or permanent brain damage may result unless treated immediately.	Thermo-regulatory system breaks down under stress and sweating stops. The body's ability to remove excess heat is almost eliminated.	Remove to cool area; soak clothing with cold water; fan body; call physician/ambulance immediately.

The temperature of the shell may be very different from the core temperature. Under cold conditions, with a core temperature maintained at approximately 37°C, the trunk may have skin temperatures of about 36°C; the thighs may be at 34°C, the upper arms and knees at 32°C, the lower arms at 28°C, while the toes and fingers may be at 25°C. These large differences in skin temperatures indicate the effectiveness of the human thermoregulatory mechanisms, particularly shunting and vasoconstriction, in the cold. They also indicate how the body must be protected by clothing to avoid chilling of body segments below an acceptable temperature. Obviously, fingers and feet need special protection in cold conditions.

The temperatures of the neck and head do not vary much. Here, the "core" is close to the shell, and very little volume between skin and core is available for vasomotor and sudomotor functions. To keep the core warm, skin temperature must be maintained at a rather constant and high level. Hence, in the cold it is of particular importance to provide external insulation through suitable clothing for head and neck, otherwise much energy would be lost through the exposed surfaces.

In the heat, the temperature throughout the body is much more constant. The core temperature is approximated in many of the external layers, with the largest differences again found in the extremities, e.g., hands and feet.

Acclimatization

Continuous or repeated exposure to hot or cold conditions brings about a gradual adjustment of body functions resulting in a better tolerance for the climatic stress and in maintenance or improvement of physical work capabilities. Acclimatization to heat is more pronounced than to cold. The improvement in heat tolerance is most clearly demonstrated by an increased sweat production, lowered skin and core temperature, and by a reduced heart rate, compared with the reactions of the unacclimatized person. The progress of acclimation is very pronounced within about a week, and full acclimatization is achieved within about two weeks. Interruption of heat exposure of just a few days reduces the lingering effects of acclimatization, which is entirely lost after about two weeks upon return to a moderate climate.

This heat acclimatization is brought about by improved control of the vascular flow, by an augmented stroke volume accompanied by a reduced heart rate, and by higher sweat production. The improvement in sudomotor action is most prominent and manifests itself not only by larger sweat volume but also by an equalization of the sweat production over time and an increase of the activities of the sweat glands of the trunk and the extremities. Perspiration on the face and the feeling of "sweating" becomes less with heat acclimation, although total sweat production may be doubled after several days of exposure to the hot environment. More volume and better regulation of sweat distribution are the primary means of the human body to bring about cooler skin temperatures, which result in dissipation of metabolic heat.

Sudomotor regulation is accompanied by and intertwined with vasomotor improvements. The reduced skin temperature (lowered through sweating) allows a redistribution of the blood flow away from the skin surfaces, which need more blood during initial exposure to heat. Acclimation re-establishes normal blood distribution within a week or two. Cardiac output must remain rather constant even during initial heat exposure, when an increase in heart rate and a reduction in stroke volume occur. Both rate and volume are reciprocally adjusted during acclimation since arterial blood pressure remains essentially unaltered. There may also be a (relatively small) change in total blood volume during acclimation, particularly an increase of plasma volume during the first phase of adjustment to heat.

While a healthy and well trained person acclimates more easily than somebody in poor physical condition, training cannot replace acclimatization. However, if physical work must be performed in a hot climate, then such work should also be included in the acclimation phase. It is of interest to note that adjustment to heat will take place whether the climate is hot and dry, or hot and humid. Furthermore, acclimatization seems to be unaffected by the type of work performed, i.e., heavy and short or moderate but continuous. It is important that during acclimation and throughout heat exposure fluid and salt losses be replaced.

Acclimatization to cold is much less pronounced; in fact there is doubt that true physiological adjustment takes place to moderate cold when appropriate clothing is worn. The first reaction of the body exposed to cold temperature is shivering — the generation of metabolic heat to counteract heat loss. Also, some changes in local blood flow are apparent. In laboratory exposure to extreme cold conditions, with little shelter offered by clothing, even hormonal and other changes have been observed. However, normally the adjustment to cold conditions is more one of proper clothing and work behavior than of pronounced changes in physiological and regulatory functions. Thus, with relative little changes in physiological functions, food intake or rate of heat production is not much, if at all, changed under "normal" working conditions in "normally cold" temperatures. It appears that with proper clothing the actual cold exposure of the body is not very severe and does not require appreciable increases in metabolism or other major adjustments in vasomotor or sudomotor systems. However, there are so-called "local" acclimatizations, particularly increased blood flow through the hands or in the face.

On the average, compared to the male the female has smaller body mass (about 80%) i.e., a smaller heat "sink" than the male; women usually have relatively more body fat and accordingly less lean body mass than men. However, their surface area is smaller, and their blood volume is smaller as well. Under heat stress, many females show somewhat lower metabolic heat production, have a higher set point; begin to sweat at higher temperatures, and may acclimate more slowly to very hot conditions; under cold stress, they have slightly colder temperatures at their (thinner) extremities but show no difference in core temperature. Altogether, there are no great differences between females and males with respect to their ability to adapt to either hot or cold climates, with women possibly at a slightly higher risk for heat exhaustion and collapse and for cold injuries to extremities. However, these slight statistical tendencies can be easily counteracted by ergonomic means and may not be obvious at all when observing only a few persons of either gender.

DESIGNING THE THERMAL ENVIRONMENT

There are many ways to generate a thermal environment that is both suitable to the physiological functions for the (acclimatized or non-acclimatized) person, and which brings about thermal comfort. The technical means to influence the climate must be seen, obviously, in interaction with the work to be performed, with the acclimatization condition of the individuals, with their clothing, and their psychological inclination to either accept given conditions or to consider them uncomfortable.

The physical conditions of the climate (humidity, air movement, temperatures) influence the cooling or heating of the body via the heat transfer functions (radiation, convection, conduction, and evaporation). These interactions are listed in Table 9-2 and must be carefully considered when designing and controlling the environment.

Table 9-2. Designing the thermal environment to increase (+) or decrease (−) body heat content by changing climate parameters.

Heat Transfer	Air Humidity		Air Movement		TEMPERATURES (as compared to skin) of					
					Air, Water		Solids		Opposing Surface	
	Dry	Moist	Fast	Calm	Hotter	Colder	Hotter	Colder	Hotter	Colder
Radiative	No direct effect		No direct effect		NA		NA		+	−
Convective	No direct effect		−	(−)	+	−	NA		NA	
Conductive	NA		NA		NA		+	−	NA	
Evaporative	−	(−)	−	(−)	−	(−)	NA		NA	

parentheses indicate that the heat loss is not as pronounced as in the corresponding condition.

Microclimate and Thermal Comfort

What is of importance to the individual is not the climate in general, the so-called macroclimate, but the climatic conditions with which one interacts directly. Every person prefers a *microclimate* that feels "comfortable" under given conditions of adaptation, clothing, and work. The suitable microclimate is highly individual and also variable. It depends on gender, as just discussed. It also depends somewhat on age, where with increasing years the muscle tonus is reduced; older persons tend to be less active and to have weaker muscles, to have a reduced caloric intake, and to start sweating at higher skin temperatures. It depends on the surface-to-volume ratio, which for example in children is much larger than in adults, and on the fat-to-lean body mass ratio.

Thermal comfort depends largely on the type and intensity of work performed. Physical work in the cold may lead to increased heat production and hence to less sensitivity to the cold environment, while in the heat hard physical work could be highly detrimental to the achievement of an energy balance. The effects of the microclimate on mental work are rather unclear, with the only sure common-sense statement that extreme climates hinder mental work.

Clothing largely affects the microclimate. The insulating value of clothing is defined in *clo* units, with 1 clo = 0.16 $°C^{-1}$ W^{-1} m^{-2}, the value of the "normal" clothing worn by a sitting subject at rest in a room at about 21°C and 50% relative humidity. Air bubbles contained in the clothing material or between clothing layers provide increased insulation, both against hot and cold environments. Permeability to fluid (sweat) and air plays a major role (Parsons 1988). Colors of the clothes are important in a heat radiating environment, such as in sunshine, with darker colors absorbing heat radiation and light ones reflecting incident energy.

Clothing also determines the surface area of exposed skin. More exposed surface areas allow better dissipation of heat in a hot environment but can lead to undercooling in the cold with heat transmitted through convection and through radiation to colder media.

Convection heat loss is increased, if the air moves swiftly along exposed skin surfaces. Therefore, with increased air velocity, body cooling becomes more pronounced. During WWII, experiments were performed on the effects of ambient temperature and air movement on the cooling of water. These physical effects were also assessed psychophysically in terms of the *wind chill* sensation at exposed human skin. Table 9-3 shows the energy loss depending on air temperature and movement but not considering humidity.

Table 9-4 shows how exposed human skin reacts to energy losses brought about by air velocity at various air temperatures. It lists the *wind chill equivalents* in degrees Celsius which reflect the effects of air velocities at various temperatures. Note that these wind chill temperatures are based on the cooling of exposed body surfaces, not on the cooling of a clothed person. Also, these numbers do not take into account air humidity. Under humid conditions, freezing of flesh may occur at wind chill values as low as 3500 kJ m^{-2} hr^{-1} (800 kcal m^{-2} hr^{-1}).

Thermocomfort is obviously also affected by acclimatization, i.e., the status of the body (and mind) of having adjusted to changed environmental conditions. A climate that was rather uncomfortable and restricted one's ability to perform physical work during the first day of exposure may be quite agreeable after two weeks. Relatedly, seasonal changes in climate, usual work, clothing, and attitude play a major role in what appears to be acceptable or not. In the summer, most people are willing to find warmer, windier, and more humid conditions comfortable than they would in the winter.

Various combinations of climate factors (temperature, humidity, air movement), can subjectively appear as similar though depending on work tasks, clothing, etc. The WBGT discussed earlier is one attempt to establish climate factors that, combined, have equivalent effects on the human. Similar approaches have been proposed throughout many decades, resulting in about two dozen different scales of "effective temperatures."

Table 9-3.   Wind chill values (energy loss through exposed skin) and their psychological correlates (adapted from MIL-HDBK 759A 1981).

Approximate Wind Chill Value		Human Sensation
kJ m^{-2}hr^{-1}	kcal m^{-2}hr^{-1}	
200	50	Hot
450	110	Warm
850	210	Pleasant
1600	400	Cool
2500	600	Very cool
3500	800	Cold
4200	1000	Very cold
4900	1200	Bitterly cold

Table 9-4. "Wind chill temperature" depending on air temperature and air movement (adapted from MIL-HDBK 759A 1981).

WIND SPEED (m s^{-1})	ACTUAL AIR TEMPERATURE (deg C)														
CALM	4	2	-1	-4	-7	-9	-12	-15	-18	-23	-29	-34	-40	-46	-51
2.2	2	-1	-4	-7	-9	-12	-15	-18	-20	-26	-32	-37	-43	-48	-57
4.5	-1	-7	-9	-12	-15	-18	-23	-26	-29	-37	-43	-51	-57	-62	-71
6.7	-4	-9	-12	-18	-21	-23	-29	-32	-34	-43	-51	-57	-65	-73	-79
8.9	-7	-12	-15	-18	-23	-26	-32	-34	-37	-45	-54	-62	-71	-79	-84
11.2	-9	-12	-18	-21	-26	-29	-34	-37	-43	-51	-59	-68	-76	-84	-93
13.4	-12	-15	-18	-23	-29	-32	-34	-40	-46	-54	-62	-71	-79	-87	-96
15.6	-12	-15	-21	-23	-29	-34	-37	-40	-46	-54	-62	-73	-82	-90	-98
17.9	-12	-18	-21	-26	-29	-34	-37	-43	-48	-56	-65	-73	-82	-90	-101
Winds above 18 m s^{-1} have little additional effect	LITTLE DANGER							INCREASING DANGER: FLESH MAY FREEZE WITHIN 1 min				GREAT DANGER: FLESH MAY FREEZE WITHIN 30 s			

The model shown in Figure 9-2 is typical and rather well accepted. This thermal index does not include radiant heat transfer but reflects how combinations of dry air temperature, humidity, and air movement affect people wearing indoor clothing and doing sedentary or light muscular work. The result is numerically equal to the temperature of still saturated air, which induces the same sensation. For example: the dashed line in Figure 9-2 indicates that the combination of a "dry" temperature of about 24°C (A) and of a "wet" temperature of about 17°C (B) generates "effective temperatures" of about 21°C at low air speed (6 m min^{-1}) while strong winds (such as 200 m min^{-1}) lower the sensation to that of nearly 17°C effective temperature (ET).

Given these many variables, it is not surprising that the same effective temperature is considered by some people to be too warm and by others as too cold. However, at truly cold temperatures much agreement on "cold" is expected, while under a very hot condition most people will consent to "hot" (see Figure 9-3).

With appropriate clothing and light work, comfortable ranges of *effective temperature* are from about 21 °C – 27°C ET in a warm climate or during the summer, and from 18°C – 24°C ET in a cool climate or during the winter. In terms of body measurements, skin temperatures in the range of 32°C – 36°C are considered comfortable, associated with core temperatures between 36.7°C – 37.1°C. Preferred ranges of relative humidity are between 30% – 70%. Deviations from these zones are uncomfortable or even intolerable, as Figure 9-4 indicates. Air temperatures at floor level and at head level should not differ by more than about 6°C. Differences in temperatures between body surfaces and side walls should not exceed approximately 10°C. Air velocity should not exceed 0.5 m/sec, at best below 0.1 m/s. Further information for the built environment, such as in offices, is contained in the ANSI-ASHRAE Standard 55, latest edition.

SUMMARY

- The body must maintain a core temperature near 37°C with little variation despite major changes in internally developed energy (heat), in external work performed, in heat energy received from a hot environment, or in heat energy lost to a cold environment.
- Heat energy may be gained from or lost to the environment by the following:
  - radiation,
  - convection,
  - conduction,

  but lost only by evaporation.
- The major avenues of the body to control heat transfer between core and skin are the efferent motor pathways (muscle tonus), sudomotor pathways (sweat production), and vasomotor pathways (control of blood flow).
- Muscular activities are the major means to control heat generation in the body. Blood flow control affects heat transfer between body core and skin.
- In a hot environment, the body tries to keep the skin hot to prevent heat gain and to achieve heat loss. Sweating is the ultimate means to cool the body surface.
- In a cold environment, the body tries to keep the skin cold to avoid heat loss.
- Acclimatization to a hot environment includes controlled blood flow to the skin, facilitated sweating, and increased stroke volume of the heart without increase in heart rate. It can be accomplished (in healthy and fit persons) in one or two weeks and be lost just as quickly.

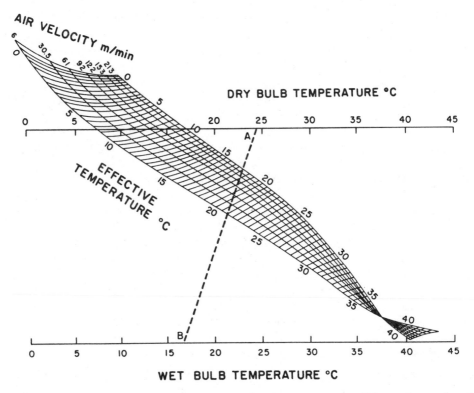

Figure 9-2. Nomogram for deriving the "effective temperature" from dry and wet bulb temperatures and from air velocity (MIL-HDBK-759A 1981).

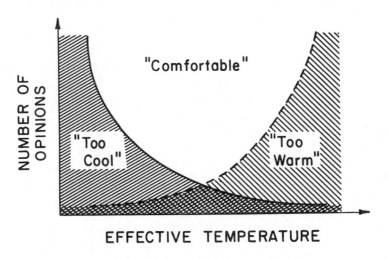

Figure 9-3. Opinions about effective temperatures.

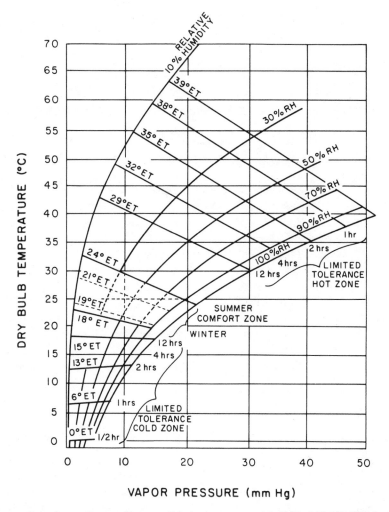

Figure 9-4. Comfort, discomfort, and tolerance zones (MIL-HDBK-759A 1981) for young healthy soldiers, expressed in degrees Celsius of effective temperature ET.

- Whether any truly physiological acclimatization to a moderate cold environment takes place is questionable, since most of the adjustment made concerns proper clothing, within which the body performs at its usual microclimate. However, blood flow to exposed surfaces and to the hands and feet is locally adapted.
- The thermal environment is determined by combinations of
  - air humidity (mostly affecting evaporation)
  - air temperature (affecting convection and evaporation)
  - air movement (affecting convection and evaporation)
  - temperature of solids in touch with the body (affecting conduction)
  - temperature of surfaces distant from the body (affecting radiation).
- The combined effects of all or some of these physical climate factors can be expressed in form of a "climate index." Various indices are in use, such as the WBGT and ET scales.

- Certain ranges of humidity, temperatures, and air velocity have been identified as "comfortable" for given tasks and clothing.

REFERENCES

ASHRAE (ed). 1981. Thermal Environment Conditions for Human Occupancy. ANSI-ASHRAE Standard 55. Atlanta, GA: American Society of Heating, Refrigerating, and Air-Conditioning Engineers.

ASHRAE (ed). 1985. Physiological Principles for Comfort and Health. In *1985 Fundamentals Handbook*. (Chapter 8). Atlanta, GA: American Society of Heating, Refrigerating, and Air-Conditioning Engineers.

Nadel, E. R. 1984. Energy Exchanges in Water. *Undersea Biomedical Research*, 11:4:149–158.

Nadel, E. R. and Horvath, S. M. 1975. Optimal Evaluation of Cold Tolerance in Man. In S. M. Horvath, S. Kondo, H. Matsui and H. Yoshimura (Eds.). *Comparative Studies on Human Adaptability of Japanese Caucasians, and Japanese Americans*, (Chapter 6A). Vol. 1. Tokyo: Japanese Committee of International Biological Program.

Spain, W. H., Ewing, W. M., and Clay, E. 1985. Knowledge of Causes, Control Aids, Prevention of Heat Stress. *Occupational Health and Safety*. 54:4:27–33.

Stolwijk, J. A. J. 1980. Partitional Calorimetry. In Assessment of Energy Metabolism in J. M. Kinney (Ed.) *Health and Desease*. (pp. 21–22). Columbus, OH: Ross Laboratories.

Stegemann, J. 1984. *Leistungsphysiologie*. Third edition. Stuttgart and New York, NY: Thieme.

MIL-HDBK-759A. 1981. Human Factors Engineering Design for Army Material. Redstone Arsenal, AL: U.S. Army Missile Command.

FURTHER READING

Astrand, P. 0. and Rodahl, K. 1986. *Textbook of Work Physiology*. Third edition. New York, NY: McGraw-Hill.

Baker, M. A. Ed. 1987. *Sex Differences in Human Performance*. Chichester: Wiley.

Kobrick, J. L. and Fine, B. J. 1983. Climate and Human Performance. In D. J. Osborne and M. M. Gruneberg (Eds.). *The Physical Environment at Work* (pp, 69–107). Chichester, Sussex: Wiley.

Stegemann, J. 1981. *Exercise Physiology*. Chicago, IL: Year Book Medical Publications, Thieme.

Chapter 10

WORK SCHEDULES AND BODY RHYTHMS

OVERVIEW

The human body changes its physiological functions throughout the 24-hour day. During waking hours, the body is prepared for physical work, while during the night sleep is normal. Attitudes and behavior also change rhythmically during the day. The diurnal rhythms can be put out of order by imposing a new set of time signals and activity-rest regimen, such as associated with shiftwork schedules. Shiftwork should be arranged to least disturb physiological, psychological, and behavioral rhythms to avoid negative health and social effects and to avoid reductions in work performance.

The Model

Daily rhythms are systems of temporal programs within the human organism. They should be left intact for continued normal functioning, both physically and psychologically, by selection of suitable work schedules.

INTRODUCTION

The human body follows a set of daily fluctuations, called circadian rhythms (from the Latin *circa,* about, and *dies,* the day; also called diurnal from the Latin *diurnus,* of the day). These are regular physiological occurrences, observable for example in body temperature, heart rate, blood pressure, and hormone excretion (see Figure 10-1).

Daily rhythms are systems of temporal programs within the human organism. They are characterized by the fact that they are manifest, well established under rigorous experimental conditions and valid by actual observation and experience. They also are characterized by their persistence under varying external conditions. Each is controlled within the body by a self-sustained "internal clock" which runs on a 24- to 25-hour cycle. Several rhythmic programs, such as core temperature, blood pressure, and "sleepiness" are coupled with each other.

However, the rhythms can be put "out of order" by external events, primarily by imposing a new set of external events and time markers on the body. The time marker is called zeitgeber (from the German *Zeit,* time; and *geber,* giver). Among the zeitgebers are daily light/darkness, true clocks, and temporally established activities, such as office hours, meal times, etc. The strengths of these zeitgebers vary.

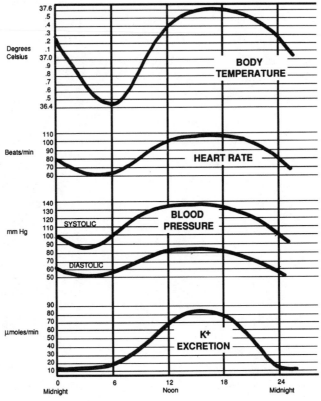

Figure 10-1. Typical Variations in Body Functions over the Day (adapted from Minors and Waterhouse 1985).

Human social behavior (the inclinations to do certain activities as well as to rest and sleep) follows fairly obvious rhythms and sequences during the day. Of course, there are other "chrono-biological" variations that one experiences: some are well documented, others postulated or mythical, such as dependency on the moon phases. So-called bio-rhythms were a fad a few decades ago: they were said to be regular waves of physiological and psychological events, starting at birth but running in different phases and phase lengths. Whenever "positive" phases of any of these rhythmic phenomena coincided, a person was believed to be under positive conditions, able to perform exceptionally well. In contrast, if "negative" phases concurred, the person was sup-posedly doing badly. Research has shown, conclusively, that no such accumulations of positive or negative phases were demonstrable; also, several or all of the supposedly existing rhythms were either myths or artifacts (Hunter and Shane, 1979; Persinger, Look, and Janes 1978).

FEMALE MENSTRUAL CYCLE

The female menstrual cycle is regulated through synchronization of the activities of the hypothalamus, pituitary, and ovary. The 28-day time period is usually divided into five phases: (1) pre-ovulatory or follicular, (2) ovulatory, (3) post-ovulatory or luteal, (4) pre-menstrual, and (5) menstrual. Main hormonal changes occur in the release of estrogen and progesterone around the 21st day of menstruation; estrogen shows a second peak at ovulation. Hormonal release is low during the pre-menstrual phase.

It has been commonly assumed that hormonal changes during the menstrual cycle have profound effects on a woman's psychological and physiological states. However, after reviewing the existing scientific work, Patkai (1985) states that the information available is fairly weak in that almost all work has been done trying to correlate certain behavioral or physiological events with the menstrual phase. But the observable events are fairly weak in their occurrence and of course an existing correlation does not necessarily indicate a causal relationship. Patkai deplores that assertations were "made by some researchers on the basis of scientifically poor or scanty data, depicting women as helpless weak victims of the ebb and flow of their hormones" (p. 88). Certainly, a balanced research position is necessary which considers the close interplay of physiological and psychological factors.

The bulk of existing research relies on self-reported changes in mood and physical complaints in the course of the menstrual cycle. Fairly little information is available on changes in arousal and on objective measures of performance. While there is neurophysiological evidence that estrogen and progesterone affect brain function, it must be considered that these two hormones have antagonistic effects on the central nervous system, with estrogen stimulating and progesterone inhibiting. Varying hormone production during the menstrual cycle may affect the capacity to perform certain tasks, but the extent to which hormones actually determine performance depends on how a decrease in total capacity may be offset by increased effort. Patkai cites a study in which secretaries showed the highest typing speed before the onset of menstruation and during the first three menstrual days. The idea of a higher effort on these days was rejected by the secretaries since they considered themselves to be working at full capacity all the time.

The occurrence of negative moods and physical complaints in the majority of women before and during menstruation is fairly well established, but the precise nature of the so-called pre-menstrual syndrome is not yet determined. According to Patkai, there is evidence that menstruation can bring about negative social behaviors, which however are mediated by social and psychological factors.

Summary: the hypothesis of reduced performance during the pre-menstrual and the menstrual phases is not well supported by objective data.

CIRCADIAN RHYTHMS

The prerequisite for human health is the maintenance of physiological variables in spite of external disturbances. This state of balanced control is called homeostasis. However, a close look at this supposedly "steady state" of the body reveals that many physiological functions are in fact not constant but show rhythmic variations. Rhythmic variation means that quantitative events (such as ups or downs in body temperature or hormone secretion) follow each other regularly. The periods of different rhythms are quite diverse, such as the heart beating about once every second, body temperature having its peak value every 24 hours, or a woman's menstrual cycle reoccurring every 28 days. While rhythms with a cycle length of 24 hours are called circadian or diurnal rhythms, those which oscillate faster than once every 24 hours are called ultradian; those which repeat less frequently, infradian.

Among the circadian rhythms, the best known physiological variables are body temperature, heart rate, blood pressure, and the excretion of potassium. Most of these variables show a high value during the day and lower values during the night, although hormones in the blood tend to be more concentrated during the night, particularly in the early morning hours. The variations during the circadian circle are fairly small, approximately ±1°C for oral temperature; many are in the range of approximately ±15% about the average, such as heart rate and diastolic blood pressure; however, some vary considerably, such as triglycerides which vary by nearly ±80% in the blood serum, while the sodium contents of the urine oscillates even more. The amount by which the variables change during the diurnal variation and the temporal locations of rhythm extremes during the day, are quite different among individuals and can change even within one person (Minors and Waterhouse 1981; Folkard and Monk 1985).

One way to observe the diurnal rhythms and to assess their effects on performance, is simply to observe the activities of a person. During daytime, a person is normally expected to be awake, active, and eating while at night sleeping and fasting. Physiological events do not exactly follow that general pattern. For example, body core temperature falls even after the person has been sleeping for several hours; it is usually lowest between 3 and 5 o'clock in the morning. Core temperature then rises quickly when one gets up. It continues to increase, with some variations, until late in the evening. Thus, body temperature is not a passive response to our regular daily behavior, such as getting up, eating meals, performing work, and doing other social activities but is self-governed.

While there exist interactions among the external activities and their zeitgebers, the underlying physiological rhythms of the body are solid, self-regulated, and remain in existence even if daily activities change. Variations in observed rhythmic events (due to exogenous influences) may occur or "mask" the internal regular fluctuations. For example, skin temperature (particularly at the extremities) increases with the onset of sleep, regardless of when this occurs. Turning the lights on increases the activity level of birds, regardless of when this occurs. Thus, skin temperature or activity level do not necessarily indicate the internal rhythm but may in fact mask it. Of course, under regular circumstances there is a well established phase coincidence between the external activity signs and the internal events. For example, during the night, the low values of physiological functions, for example, core temperature and heart rate are primarily due to the diurnal rhythm of the body; however, they are further helped by nighttime inactivity and fasting. During the day, peak activity usually coincides with high values of the internal functions. Thus, normally, the observed diurnal rhythm is the result of internal (endogenous) and external (exogenous) events which concur. If that balance of concurrent events is disturbed, consequences in health or performance may become apparent.

When a person is completely isolated from external factors (zeitgebers), including regular activities, the internal body rhythms are "running free." This means that the circadian rhythms are free from external time cues and are only internally controlled. Many experiments have consistently shown that circadian rhythms persist when running free, but their time periods are slightly different from the regular 24-hour duration: most rhythms "run free" at about 25 hours, some take longer. Since the earth continues to rotate at 24 hours, this experience indicates that body rhythms are independent from external stimuli and follow their own built-in clocks. However, if a person is put again under daily (24 hour) zeitgebers and activities, the internal rhythms resume their 24-hour cycles.

Oscillatory Control

The phenomena of human rhythms often have been explained by an oscillator model of the human circadian system. This assumes that various overt rhythms are jointly controlled by a few basic oscillators which, however, may have different controlling power.

The basic oscillators are in turn controlled by external stimuli, and they also influence each other. If their intrinsic periods are close together, they synchronize. This internal coordination falls apart, for example, when artificial zeitgebers occur within the entrainment range of one oscillator but outside of another. Rhythms controlled by the first oscillator will remain entrained, but those controlled by the other oscillator will begin to run free: internal desynchronization takes place. For example, the sleep/wake cycle may remain entrained, such as at 26 hours, while the temperature rhythm may run freely at a period of 25 hours (Wever 1985).

To investigate the constancy or temporal isolation of rhythms, two types of experiments have been performed. One uses the absence of any natural or artificial time cues to evaluate the purely internal control. Other experiments use the influence of artificial zeitgebers of various types to evaluate the effects of internal and external factors.

Under constant experimental conditions, i.e., without zeitgebers, human circadian rhythms are free-running at about 25-hour periods. This has shown to be true both for isolated individuals as well as for groups of subjects, although some intraindividual as well as a somewhat smaller interindividual variability exists (Wever 1985). Since there is no longer synchronization between the free-running cycles and the 24-hour day, this condition is also called "desynchronization" of internal functions from the 24-hour zeitgebers.

Manipulation of zeitgebers allows laboratory simulation of jet lag or shiftwork. Experiments with artificial zeitgebers have shown that these play a major role in "entraining" or synchronizing the internal rhythms so that they follow the periodic time cues. Synchronization of the internal rhythms to time events has been demonstrated to be possible with cycle durations between 23 to 27 hours. (At shorter or longer periods of time cues, the circadian rhythms are free-running, though often not completely independent of the time cues.) Most researchers have concluded that it is easier to set one's internal clocks "forward," such as it occurs in the spring when "daylight-savings time" is introduced in North America and in Europe. However, a recent study on shiftwork did not indicate a significant beneficial effect of forward rotation as compared to backward rotation (Duchon, Wagner, and Keran 1989).

Individual Differences

Experimentation has shown that some people have consistently shorter (or longer) free-running periods than others. For example, those who have short periods are likely to be "morning types" while those with longer internal rhythms are probably "evening types." It appears that females have, on the average, a free-running period that is about 30 minutes shorter than that of males. This suggests that females may be more prone to rhythm disorders than males (Wever 1985).

There is an interaction between aging and circadian rhythms. Rhythm amplitudes usually are reduced with increasing age. This is particularly obvious for body temperature. The temperature rhythm also appears to be advanced relative to the mid-sleep period, which agrees with the finding of a shift towards morning with increasing age. Also, if the oscillatory controls lose some of their power with increasing age, as appears to be true, this would indicate a greater susceptibility to rhythm disturbances with increasing age (Kerkhof 1985).

Daily Performance Rhythms

Given the systematic changes in physiological functions during the day, one expects corresponding changes in mood and performance. Of course, attitudes and work habits are also, and often strongly, affected by the daily organization of getting up, working, eating, relaxing, and going to bed. Experimentally, one can separate the effects of internal circadian rhythms and of external daily organization. For practical purposes, one wants to look at the results (e.g., as they affect performance) of the internal and external factors combined.

Early in this century, it was thought that the morning hours would be best for mental activities, with the afternoon more suitable for motoric work. On the other hand, "fatigue" arising from work already performed was believed to reduce performance over the course of the day. For simple mental work, such as recording numbers, it was observed that performance showed a pronounced reduction early in the afternoon. This was labeled the "post-lunch dip." However, this reduction in performance was not paralleled by a similar change in physiological functions; for example, body temperature remains fairly unchanged at that period of the day. Hence, it was postulated that the interruption of the activities by a noon meal and the following digestive activities of the body, would bring about this often observed reduction in performance.

Such post-lunch dips are found mostly with activities that can be related to the psychological "arousal" level. Lunch time might bring about an increased lassitude, a status of deactivation, paralleled by or associated with increased blood glucose and pulse rate, possibly the results of food ingestion. In this case, the post-lunch dip appears to be caused by the exogenous "masking" effect of food intake, rather than by endogenous circadian effects.

However, in other activities, primarily those with medium to heavy physical work, no such dip has been found after lunch (except when the food and beverage ingestion was very heavy and if true physiological fatigue had been built up during the pre-lunch activities).

Summary: many different activities may be performed during the day. Some of these strongly follow a circadian rhythm, some less; on many, exogenous masking effects may be more pronounced than on others. For example, information processing in the brain (including immediate or short-term memory demands), mental arithmetic activities, or visual searches, may be strongly affected by personality, or by the length of the activity and by motivation. Thus, it appears that one cannot make "normative" statements about diurnal performance variations or abilities, during regular working hours.

SLEEP

Two millennia ago, Aristotles thought that during wakefulness some substance ("warm vapors") in the brain built up which needed to be dissipated during sleep. In the 19th Century, there were two opposing schools of thought: one that sleep was caused by some "congestion of the brain by blood," the other that blood was "drawn away from the brain." Also, "behavioral" theories were common in the 19th century, such that sleep was the result of an absence of external stimulation or that sleep was not a passive response but an activity to avoid fatigue from occurring. Early in the 20th century, it was thought that various sleep-inducing substances accumulated in the brain, an idea taken up again in the 1960s. In the 1930s and 1940s, various "neural inhibition" theories were discussed, including sleep-inducing "centers," such as arousal centers in the reticular formation of the brain (Horne 1988).

"Restorative" theories about the function of sleep focus on various types of "recovery" from the wear and tear of wakefulness. Alternative theories reject this idea and claim that sleep is not restorative but simply a form of instinct or "non-behavior" to occupy the unproductive hours of darkness; through relative immobility of the body, sleep may be a means to conserve energy (Horne 1988). Horne believes that the three aspects of sleep function, i.e., restoration, energy conservation, and occupying time, all explain certain characteristics of sleep but neither of them completely or sufficiently.

According to Horne, it is convenient to consider that the regulation of alertness, wakefulness, sleepiness, sleep, and of many physiological functions is under the control of two "central clocks" of the body. One controls sleep and wakefulness, the other physiological functions, such as body temperature. Under normal conditions, the internal clocks are linked together so that body temperature and other physiological activities increase during wakefulness and decline during sleep. However, this congruence of the two rhythms may be disturbed, for instance by night shiftwork, where one must be active during nighttime and sleep during the day. As such patterns continue, the physiological clocks adjust to the external requirements of the new sleep/wake regimen. This means that the formerly well-established physiological rhythm flattens out and, within a period of about two weeks, re-establishes itself according to the new sleep/wake schedule.

Sleep Phases

The brain and muscles are the human organs that show the largest changes from sleep to wakefulness: their activities can be observed by electrical means.

To observe human sleep, electrodes are attached to the surface of the scalp. They pick up electrical activities of the cortex, which is also called "encephalon" because it wraps around the inner brain. Thus, the measuring technique is named electro-encephalo-graphy, EEG. The EEG signals provide information about the activities of the brain. It is also common to record the electrical activities associated with the muscles that move the eyes and those in the chin and neck regions. The electrical recording of muscle (greek, *myo*) activities is called electro-myo-graphy, EMG. [Currently, in sleep research EMG signals are not analyzed to the extent common in biomechanics or physiology (Basmajian and deLuca 1985).]

EEG signals can be described in terms of amplitude and frequency. The amplitude is measured in microvolts, with the amplitude rising as consciousness falls from alert wakefulness through drowsiness to deep sleep. EEG frequency is measured in hertz; the frequencies observed in human EEG range from 0.5 Hz to 25 Hz. Frequencies above 15 Hz are called "fast waves," frequencies under 3.5 Hz "slow waves." Frequency falls as sleep deepens; "slow wave sleep" (SWS) is of particular interest to sleep researchers.

Certain frequency bands have been given Greek letters. The main divisions are the following:

Beta, above 15 Hz. Such fast waves of low amplitude (under 10 microvolt) occur when the cerebrum is alert or even anxious.

Alpha, between 8 and 11 Hz. These frequencies occur during relaxed wakefulness when there is little information input to the eyes, particularly when they are closed.

Theta, between 3.5 and 7.5 Hz. These frequencies are associated with drowsiness and light sleep.

Delta, slow waves under 3.5 Hz. These are waves of large amplitude, often over 100 microvolt, and occur more often as sleep becomes deeper.

Also, certain occurrences in the EEG waves have been labelled, such as vertices, spindles, and complexes, which appear regularly, associated with sleep characteristics.

The importance placed upon EEG and EMG events by sleep researchers has been changing over decades. Currently, EMG outputs of the eye muscles are most often used as the main determiner: sleep is divided into periods associated with rapid eye movements, REM, and those without, non-REM. Non-REM conditions are further subdivided in 4 stages according to their associated EEG characteristics. Table 10-1 lists these.

The REM sleep phase is accompanied by irregular breathing and heart rate, and by low voltage, fast brain activities visible in the EEG. In the other sleep phase, non-REM, regular and slow breathing and heart rates occur, and the EEG activity is slow but shows high voltage. These phases change cyclicly during the sleep and are probably co-organized between diverse brain regions, thus involving two or more oscillators. The REM/non-REM cycles occur in roughly 1.5-hour timings, an ultradian (shorter-than-daily) rhythm. However, this duration of 1.5 hours has large within- and between-subjects variability and appears to shorten in the course of a night's sleep, accompanied by a relative lengthening of the REM portion.

In 1963, Kleitman (according to Lavie 1985) suggested the existence of a Basic Rest-Activity Cycle (BRAC) running throughout the day. Such BRAC would explain the observation that many renal, gastric, eating, behavioral as well as mental activities follow roughly a 1.5-hour cyclicity. In particular, it would explain fluctuations in alertness and arousal, inattention, day dreaming, and sleepiness during the waking hours. Thus, Kleitman (according to Lavie) proposed that suitable jobs be organized into working periods and breaks so that each work unit lasts approximately 1.5 hours.

Table 10-1. Sleep Stages (adapted from Horne, 1988; and Rechtschaffen and Kales, 1968).

Condition	Muscle EMG	Brain EEG	Sleep Stage	Average Percent of Total Sleep Time
awake	active	active, alpha & beta	0	---
drowsy, transitional "light sleep"	eyelids open and close, eyes roll	theta, loss of alpha, vertex sharp waves	1, non-REM	5
"true" sleep		theta, few delta, sleep spindles, K-complexes	2, non-REM	45
transitional "true" sleep		more delta, SWS (< 3.5 Hz)	3, non-REM	7
deep "true" sleep		predominant delta SWS (< 3.5 Hz)	4, non-REM	13
sleeping	rapid eye movements, other muscles relaxed	alert, much dreaming, alpha and beta	REM	30

Sleep Loss and Tiredness

If a person does not get the usual amount of sleep, the apparent result is tiredness, and the obvious cure is to get more sleep. Shiftworkers are particularly subject to this problem, when they have to make up for sleep lost during their regular resting hours in daytime. Figure 10-2 shows the effects of sleep loss on body temperature — the temperature is raised during the night and morning but keeps its phase.

It is of some interest to note that it is not entirely clear why humans (or animals) need sleep. There is the general opinion that sleep has recuperative benefits, allowing some sort of restitution or repair of tissue or brain following the "wear and tear" of wakefulness. However, what is meant by restitution or repair is usually not clearly expressed, nor fully understood. Certainly, sleep is accompanied by rest and, to a large extent, by energy conservation. But, a human can attain similar relaxation during wakefulness, when not forced to be active. Regarding restitution of the body, it has been found that certain hormones are released more during sleep than during wakefulness; prominent among these hormones is the human growth hormone. However, few such "positive" sleep events have been noted. Protein in tissues is continuously broken down into its amino-acid building blocks or reconstituted from recent food intake. If such breakdown were excessive during wakefulness, then the rate of synthesis should be especially high during sleep. This was not found to be the case: in fact, during sleep, protein synthesis is low and breakdown increased, leading to the loss of protein through dissolution, instead of an increase through restitution.

Many experiments have failed to show restitutive physiological effects of sleep; in fact, even moderate sleep deprivation has little physiological effect (though there are clear signs of impairment of the central nervous system), as discussed by Horne. For example, sleep deprivation does not impair muscle restitution or the physiological ability to perform physical work. Apparent reductions in physical exercise capability, owing to sleep deprivation (such as reported by Froeberg 1985) may be mostly due to reduced psychological motivation rather than to a decrease in physiological capabilities. The effects of sleep deprivation on body functions are not clear but may be less consequential than often believed.

The lack of experimental findings regarding the physical benefits of sleep are somewhat surprising because they do not coincide with common experience. After physically strenuous work or exercise, tired and hurting muscles are obviously recuperated after a good night's sleep. Would physiological restoration processes such as the tissue rebuilding or metabolizing of lactic acid, which apparently take place during the sleep, really occur as well during waking rest?

In contrast, the restorative benefits of sleep to the brain are fairly well researched. Two or more nights of sleep deprivation bring about psychological performance detriments, particularly reduced motivation to perform (but apparently not a reduction of the inherent cognitive capacity), behavioral irritability, suspiciousness, speech slurring, and other performance reductions. However, while these changes indicate some CNS impairment owing to sleep deprivation (a need for the brain to sleep), Horne (1985, 1988) states that they are not as extensive as one might expect a person needed eight hours of sleep per day for brain restitution. Even though one feels tired, mental performance is still rather normal after up to two days of sleep deprivation on stimulating and motivating tasks; however, boring tasks show performance reduction. (All task performance is reduced after more than two nights of sleep deprivation.) It is of some interest to note that the performance levels are lower during nighttime activities, when the body and brain usually rest.

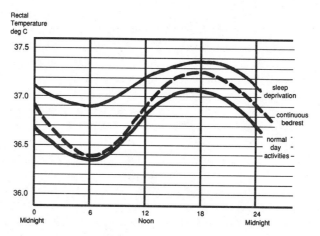

Figure 10-2. Changes in Body Temperature Associated with Bedrest, Normal Activities, and Sleep Deprivation (schematically from Colligan and Tepas 1986).

Horne (1985, 1988) speculates that, during normal mental tasks, there is an over-capacity of cerebral neural networks since many are not fully used. However, during sleep deprivation (possibly as a result of missing "restitution") the extra circuits become used, first without overt effect on performance. With increasing deprivation, more circuits become impaired, so that all available circuits finally become used up and performance drops.

Within the human body, only the brain assumes a physiological state during sleep which is unique to sleep and cannot be attained during wakefulness. While, for example, muscles can rest during relaxed wakefulness, the cerebrum remains in a condition of "quiet readiness," prepared to act on sensory input, without diminution in responsiveness (Horne, 1985, 1988). Only during sleep do cerebral functions show marked increases in thresholds of responsiveness to sensory input. In the deep sleep stages associated with slow wave non-REM sleep, the cerebrum is apparently functionally disconnected from subcortical mechanisms. Regardless of whether the cerebrum needs off-line recovery during sleep from the demands of waking activities or whether it simply disconnects and withdraws, the fact remains undisputed that the brain needs sleep to restitute, a process that cannot take place sufficiently during waking relaxation (Horne, 1985, 1988).

Horne (1985) argues that apparently not all human sleep is essential for brain restitution. For example, after sleep deprivation, usually not all lost sleep is reclaimed; long-term studies with volunteer subjects have shown that a sleep period that is 1 to 2 hours shorter than usual can be endured for many months without any consequences. It seems that the first 5 to 6 hours of regular sleep (which happen to contain most of the slow-wave non-REM sleep and at least half the REM sleep) are obligatory to retain psychological performance at normal level, but that more sleep, called "facultative" or "optional," mostly serves to "occupy unproductive hours of darkness" (Horne 1985, p. 54), with dreams (mostly in REM but not totally confined to it) the "cinema of the mind" (Horne 1988, p. 313).

Normal Sleep Requirements

While there are, as usual, variations among individuals, certain age groups in the western world show rather regular sleeping hours. For example, young adults sleep, on the average, 7.5 hours (with a standard deviation of about 1 hour). Some people are well rested after 6.5 hours of sleep, or less, while others take habitually 8.5 hours and more. The amount of slow-wave sleep in both short and long sleepers is about the same, but the amounts of REM and non-REM sleep periods differ considerably. Individuals naturally sleeping less than 3.5 hours are very rare among middle-aged people; no true non-sleepers have ever been found among otherwise healthy persons (Horne 1985; 1988).

If people can sleep for just a few hours per day, many are able to keep up their performance levels even if the attained total sleep time is shorter than normal. The limit seems to lie around to 5 hours of sleep per day, with even shorter periods still being somewhat beneficial.

People can learn to extend their sleeping hours, such as taking another hour or more after the alarm rang. The regular sleep duration also can be extended by day-time napping.

Summary: the body's physical need for restitutive sleep is not clearly understood. Some sleep, particularly of the SWS type, may be the only state where the cerebrum can obtain some form of off-line recovery. No organ other than the brain shows a physiological state during sleep which is unique to sleep and cannot be attained during wakefulness. Thus, it appears that the restorative effects of sleep are mostly beneficial for the brain, less so (and not well understood) for the rest of the body. "Moderate" loss of sleep is not very consequential for performance.

PROLONGED HOURS OF WORK AND SLEEP DEPRIVATION

There are conditions in which persons must continue working for long periods of time, such as a full day or longer. This not only means working for long periods of time without interruption but also encompasses deprivation of sleep. Hence, negative results of such long working spells are partly a function of the long work itself and partly of "sleepiness." A discussion of this topic is difficult because different types of tasks may be performed, because motivation of the workers can play an important role regarding performance, and wakefulness or sleepiness appears in cycles during the day.

Tasks

Performing different types of work is affected differently by long periods of work.

A task that must be performed uninterruptedly for periods of a half hour or longer is more easily affected negatively by sleep loss than a short work task. Also, if such a task must be replicated, performance is likely to become worse with each successive repetition. Monotonous tasks are highly affected by sleep deprivation, but a task new to the operator is less affected by sleep loss. On the other hand, a complex task can be more affected than a simple one. Thus, both monotonous and complex tasks should not be requested from persons over long periods of time, particularly if associated with sleepiness.

Tasks which are paced by the work itself deteriorate more with sleepiness than self-paced tasks. Accuracy in performing a job may be still quite good even after losing sleep, but it takes longer to perform the job. Froeberg (1985), from whose work the statements above are taken, also found that a task that is interesting and appealing, even if it does include complex decision making, can be performed rather well even over long periods of time. But if the task is disliked and unappealing, decision making is prolonged. Memory also degrades when people are required to stay awake for long periods of time. Such memory deprivation appears to be occurring both on long-stored memory information and on short-term memory.

Summary: performance of "mental tasks" which take more than 30 minutes deteriorates if they are low in novelty, interest, and incentive, or if they are high in complexity. However, tasks that require decision making, problem solving or concept information which are highly interesting or rewarding, are resistant to deterioration.

Incurring Performance Deprivation and Recovering From It

Many performance functions are lowered after one night without sleep. The deterioration becomes more pronounced after two, three, or four nights of sleep deprivation. After missing four nights, very few people are able to stay awake and to perform even if their motivation is very high (Froeberg 1985). The following discussion assumes sleep deprivation of at least one night.

With increasing time at work, so-called "microsleeps" occur more frequently. The subject falls asleep for a few seconds, but these short periods (even if frequent) do not have much recuperative value because the subject still feels sleepy and performance still degrades. Another commonly observed event during long working times coupled with lack of sleep are periods of "no performance," also known as "lapses" or "gaps." These are short periods of reduced arousal or even of light sleep.

Naps lasting 1 to 2 hours improve subsequent performance. However, if a person is awakened from napping during a deep-sleep phase, "sleep inertia" with low performance can appear which may last up to 30 minutes. Temporal placing of a nap may have differing effects: for example, the common early afternoon nap has surprisingly little effect on performance of subsequent work. On the other hand, naps of at least 2 hours taken in the late evening or during nighttime, when the diurnal rhythm is falling, have positive effects lasting several hours, provided that the amount of sleep loss incurred until this moment is moderate, such as one night without sleep (Gillberg, 1985; Rogers, Spencer, Stone, and Nicholson, 1989).

It appears that the scientific findings about the usefulness of "naps" (for people who missed at least one night's sleep) do not concur with common experience. Many people take short naps, particularly after lunch. They claim that those 5 to 15 minutes of rest are highly helpful, for some almost necessary, to be "ready and fit" for the continued work. Perhaps the recuperative effects are too subtle, too much an interaction between physiological, psychological, and habitual traits, to be easily demonstrated in a scientific experiment.

If long working periods are unavoidable, one may try to have workers perform physical exercises when the work consists of mainly mental activities. However, this is not a sure way to prevent performance deprivation. Also, "white noise" may improve performance slightly. "Stirring" music may help. Drugs, particularly amphetamines, can restore performance to a normal level even when given after 3 nights without sleep (Froeberg 1985).

Recovery from sleep deprivation is quite fast. A full night's sleep, undisturbed, probably lasting several hours longer than usual, restores performance efficiency almost fully.

Summary: task performance is influenced by three factors: the internal diurnal rhythm of the body, the external daily organization of work activities, and subjective motivation and interest in the work. Each factor can govern, influence, or mask the effects of the others on task performance. Physical and mental performance capabilities deteriorate during long-continued work, accompanied by sleep loss. The performance decrement depends mainly on the task characteristics and on the motivation of the person working. Task duration, monotony, complexity, and repetitiveness have particularly negative effects, while exceptionally high motivation may prevent any performance decrement. Performance is particularly low during "low" periods of the circadian cycle such as in the early morning hours. Short naps during the work have some beneficial effect on performance. One long night's sleep usually restores performance to a normal level, even after extensive sleep deprivation.

SHIFTWORK

One speaks of shiftwork if two or more persons or teams of persons, work in sequence at the same workplace. Often, each worker's shift is repeated, in the same pattern, over a number of days. For the individual, shiftwork means attending the same workplace regularly at the same time ("continuous" shiftwork) or at varying times (discontinuous, rotating shiftwork).

The Development of Shiftwork

Shiftwork is not new. In Ancient Rome, deliveries were to be done, by decree, at night to relieve street congestion. Bakers have been working habitually through the late night hours. Soldiers and firefighters always have been used to night shifts. Thus, the seemingly modern requirement to be on duty at all hours of the day is not new. Though estimates vary and depend on the definition of "shiftwork," probably about 25% of all workers in developed countries are on some kind of shift schedule (Monk and Tepas 1985).

With the advent of industrialization, long working days became common with teams of workers relaying each other to maintain blast furnaces, rolling mills, glass works, and other workplaces where continuous operation was desired. Covering the 24-hour period with either two 12-hour workshifts or with three 8-hour workshifts, became a common practice. Technological or economical concerns have made shiftwork generally accepted in many industries, trades, and services, even in developing countries (Kogi 1985).

Since the industrial revolution, when 12-hour shifts were common, drastic changes in work systems have occurred. In the first part of the 20th century, the then common 6-day workweeks with 10-hour shifts were shortened. Today, many work systems use the 8-hour per day/5 workdays per week arrangement, which was introduced in many countries in the 1960s. In general, the number of days worked per week were reduced, usually to allow two weekend days to be free; also, the number of hours worked per day was reduced. At this writing, many employees in West Germany, particularly if working for the government, leave their workplaces early on Friday afternoon to return after a long weekend on Monday morning. In 1988, the German Secretary of Labor proposed a system with a "compressed workweek" in which one works only 4 days a week but 9 hours per day. Among other features, such systems may require 4 or more work crews. Thus, as seen by the individual worker, there appears to be a trend towards new forms of discontinuous systems, such as weekly rotations of alternating day and night shifts and 3-shift systems that cover less than 24 hours. Irregular rotation appears to be increasingly used, as are 6 or 7 consecutive similar shifts. Furthermore, working "overtime" has become a fairly regular feature in many employment situations.

Finally, it should not be overlooked that many alternate work systems have been used; for example, farmers used to start work at dawn and stop at dusk. Peddlers, politicians, professors, and others have always tried to set their own work schedules, often highly variable to suit specific circumstances.

Shift Systems

Shiftwork is different from "normal" day work in such that work is performed regularly during times other than morning and afternoon; and/or at a given workplace, more than that one shift is worked during the 24-hour day. A shift may be shorter or longer than or of the same duration as an 8-hour work period.

There are many diverse shift systems. For convenience they can be classified into several basic patterns, but any given shift system may comprise aspects of several patterns. Kogi (1985) lists four particularly important features of shift systems: whether a shift extends into hours that would normally be spent asleep; worked throughout the entire 7-day week; or whether they include days of rest, such as free weekend; into how many shifts the daily work hours are divided, i.e., are there two, three, or more shifts per day; and whether the shift crews rotate or work the same shifts permanently. All of these aspects, shown in Figure 10-3, are of particular concern with respect to the welfare of the shiftworker, the work performance, and the organizational scheduling.

Other identifiers of shifts and shift patterns are the starting and ending time of a shift; the number of workdays in each week; the hours of work in each week; the number of shift teams; the number of holidays per week or per rotation cycle; the number of

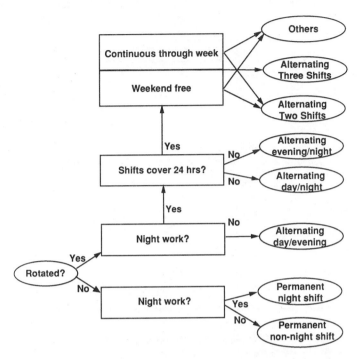

Figure 10-3. Flow Chart of Key Features of Shift Systems (adapted from Kogi 1985). Note that other shift attributes are possible.

consecutive days on the same shift, which may be a fixed or variable number; and the schedule by which an individual worker either works or has a free day, or free days (Kogi 1985).

In terms of organizing the schedule, it is easiest to set up a permanent or weekly rotation schedule. Several such solutions are shown in Table 10-2.

Many examples of work schedules are presented and discussed in the literature. (See, e.g., Colligan, and Tepas 1986; Colquhoun 1985; Eastment Kodak Company, 1986; Folkhard and Monk 1985; Johnson, Tepas, Colquhoun, and Colligan 1981; Tepas and Monk 1986.) In most systems used today, the same shift is being worked for 5 days, usually followed by 2 free days during the weekend. This regimen does not cover, however, evenly all 21 shift periods of the week, thus additional crews are needed to work on weekends or under other "odd" arrangements. If one uses three shifts a day with four teams, the shift system (for one team) is 1-1-2-2-3-3-0-0 with a 6/2 work/free day ratio and a cycle length of eight days; this is known as the "metropolitan rotation." The "continental rotation," which also assumes three shifts per day and four crews, has the sequence 1-1-2-2-3-3-3, 0-0-1-1-2-2-2, 3-3-0-0-1-1-1, 2-2-3-3-0-0-0; its work/free day ratio is 21/7, its cycle length exactly 4 weeks.

The ratio of work days versus free days in a complete cycle is an important characteristic of any shift system. Table 10-3 presents a number of other features that describe different shift systems.

Table 10-2. Examples of 5-Workdays-Per-Week Shift Systems (adapted from Kogi 1985).

System	WorkDays/Free Days	Shift Sequence
Permanent Day Shift	5/2	1-1-1-1-1-0-0, 1-1-1-1-1-0-0,
Permanent Evening Shift	5/2	2-2-2-2-2-0-0, 2-2-2-2-2-0-0,
Permanent Night Shift	5/2	3-3-3-3-3-0-0, 3-3-3-3-3-0-0,
Rotations:		
Alternating Day-Evening	10/4	1-1-1-1-1-0-0, 2-2-2-2-2-0-0.
Alternating Day-Night	10/4	1-1-1-1-1-0-0, 3-3-3-3-3-0-0,
Alternating Day-Evening-Night	15/6	1-1-1-1-1-0-0, 2-2-2-2-2-0-0,
		3-3-3-3-3-0-0 (forward rotation)
		or
		1-1-1-1-1-0-0, 3-3-3-3-3-0-0,
		2-2-2-2-2-0-0 (backward rotation)

Legend

1 represents day shift, 2 evening shift, 3 night shift, 0 free day,
  i.e., without scheduled shift.

Table 10-3. Characteristics of Shift Arrangements (Kogi 1985).

Cycle Length	$C = W + F$	
Free days per year	$D = 365\, F\, (W + F)^{-1}$	
Number of days worked before the same set of shifts re-occurs on the same days of the week	$R = C = W + F$	if $(W + F)$ is multiple of 7
	$R = 7(W + F)$	if $(W + F)$ is not a multiple of 7

Legend: W represents Work day, F Free day

A recent trend has been towards "flextime," meaning a somewhat "flexible" arrangement of work hours during the day. Instead of called flextime, this arrangement might be better described as "sliding time" (akin to the German "Gleit-Zeit") since it allows the employee to distribute the prescribed number of working hours per shift (for example, 8) over a longer block of time (10 hours) but so that a "core" time (of, say, 6 hours) is covered during which all workers must be present. Thus, one can "slide" or "float" the working time across the core time such the start of work is at any time before and the end of work anytime after the core. Flextime is often, but not necessarily combined with "compressed workweeks", where the regular number of hours of work per week (such as 40) is compressed from 5 days a week into 4.5 or even fewer days (Hurrell and Colligan 1985). Table 10-4 lists potential advantages and disadvantages of flextime.

Table 10-4. Potential advantages and disadvantages of Flextime (adapted from Tepas 1985).

Potential advantages
Generally appealing
Flexible work times with no loss in base pay
Increased day-to-day flexibility for free time
Reduces commuting problems and costs
Workforce size can adjust to short-term fluctuations in demand
Less fatigued workers
Reduces job dissatisfaction/increased job satisfaction
Increases democracy in the workforce
Recognition and utilization of employee's individual differences
Reduces tardiness
Reduces absenteeism
Reduces employee turnover
Increases production rates
Better opportunities to hire skilled worked in tight labor markets

Potential disadvantages
Irregularity in workhours produced by short-term changes in demand
Difficulty covering some jobs at all required times
Difficulty in scheduling meetings or training sessions
Poorer communication within the organization
Poorer communication with other organizations
Increases energy and maintenance costs
Increases buffer stock for assembly-line operations
More sophisticated planning, organization and control
Reduces quantity or quality of services to the public
Requires special time-recording
Additional supervisory personnel
Extension of health and food service hours

Compressed Workweeks

One speaks of a "compressed workweek" if more than the usual 8 hours per day are worked, so that the common 40 hours of work per week are performed in only 4 or even 3 days per week, thus allowing the worker to have 3 or 4 "free days" each week. This is apparently an attractive idea for many persons: it reduces the number of trips to and from work, and there are fewer "setups" and "close-downs" at work. However, there are concerns about increased fatigue due to long workdays and reduced performance and safety.

The type of work to be performed is a major determiner of whether or not compressed workweeks can and should be used. Thus, long working days have been mostly used in cases where one waits or is on "standby" much of the shift, such as firefighters. Also, activities that require only few or small physical efforts, that are diverse and interesting, yet fall into routines, have been done in long shifts. Examples are nursing, clerical work, administrative work, technical maintenance, computer supply operations, and supervision of automated processes. Little experience exists from long shifts that include manufacturing, assembly, machine operations, and other physically intensive jobs. Furthermore, information has been mostly gathered from subjective statements of employees, some limited psychological test batteries, and by scrutinizing performance and safety records in industry.

The results are contradictory, spotty, and apparently depend much on the given work conditions. In some cases, production and performance are high shortly after introduction of a compressed workweek but fall off after prolonged periods on such a schedule. However, other observations have not shown this trend. The people involved often indicate significantly increased satisfaction, which may be due more to easily and better arranged leisure time rather than to improvements of the work (Lateck and Foster, 1985). In one experimental field study, health employees changed from a 5-day/40-hour workweek to a 4-day/40-hour schedule for 4 months and then returned to the original schedule. In general, the effectiveness of the organization was somewhat improved by the compressed schedule; employees were more satisfied, and their personal activities were improved. However, none of these positive effects were strong. The return to the original 5-day workweek was not welcomed by the participants (Dunham, Pierce, and Castaneda, 1987).

Working very long workshifts, such as 12 hours, is likely to introduce drowsiness and some reduction in cognitive abilities, motor skills, and generally in performance during the course of each workshift as the workweek progresses. It appears that there is a potential for "careless shortcuts to completion of a job" by a fatigued worker, and that work practices may be less safe in tasks "that are tedious because of high cognitive or information-processing demands, or those with extensive repetition" (Rosa and Colligan, 1988, p. 315).

Table 10-5 lists potential advantages and disadvantages of compressed workweeks.

Which are suitable shift systems?

The human is used to daylight activity, with the night used for rest. This appears to be an inherent feature, as it is governed by the internal clocks of diurnal rhythms. Night work, then, appears to be "unnatural"; however, this does not necessarily mean that it is harmful. But it appears that working at night generates "stress" which may be light or severe, depending on the circumstances and the person.

Table 10-5. Potential advantages and disadvantages of compressed workweeks (adapted from Tepas 1985).

Potential advantages
Generally appealing
Increases possibilities for multi-day off-the-job activities
Reduces commuting problems and costs
More time per day for scheduling meetings or training sessions
Fewer start-up and/or warm-up expenses
Increases production rates
Improvement in the quantity or quality of services to the public
Better opportunities to hire skilled workers in tight labor markets
Potential disadvantages
Decrements in job performance due to long work hours, or to 'moonlighting' on "free" days
Overtime pay required
More fatigued workers
Increases tardiness and "leaving work early"
Increases absenteeism
Increases employee turnover
Increases on-the-job and off-the-job accidents
Decreases production rates
Scheduling problems if the organization operations are longer than the workweek
Difficulty in scheduling child care and family life during the workweek
Contrary to traditional objectives of labor unions
Increases energy and maintenance costs

Organizational criteria by which to judge the suitability of shift systems include the number of shifts per day, the length of every shift, or the times of the day during which there is no work done; the coverage of the week by shifts; shiftwork on holidays; etc. These "independent variables" have been discussed earlier.

Among the "dependent variables" is the performance of workers on shift schedules. Is the same output to be expected regardless of the time of work during the 24-hour day? Are specific activities better performed at certain shifts? Does change in shiftwork scheduling affect the shiftworker's output? Is work on certain shifts likely to show accidents?

Another "dependent variable" is the health of the shiftworker. Do certain shift regimes affect physiological or psychological "well-being"? For example: does the inability to sleep at night, when a night shift needs to be worked, affect the worker's health? What is the effect of shiftwork on "social" interactions with family, friends, and the society in general?

Thus, there are several factors of shiftwork: health and well-being, performance and accidents, psychological and social aspects. All interact with each other but not always in the same direction. Thus, conclusions for selecting a suitable work regimen depend on the given conditions.

Apart from organizational ease, work performance, well-being, and social interaction are important in the selection of a suitable shift regimen.

Regarding health and well-being, mostly in the physical sense, it is obvious that a strong circadian system exists within the body which is remarkably resistant to sudden large changes in routine. This internal system being so stable, theoretical findings, common sense and personal experiences suggest that the normal synchrony of behavior in terms of rest and activity sequences should be maintained as well as possible. Thus, work schedules should be arranged in accordance with the internal system or, if this is impossible (for example, if night work is necessary) to disturb the internal cycles as little as possible. One of the logical conclusions from this consideration is that work activities, which contradict the internal rhythm, should be kept as short as feasible so that one can return to the "normal" cyclicity as quickly as possible. For example, one should schedule single night shifts, interspaced by normal workdays, instead of requiring a worker to do a series of night shifts as shown in Table 10-2. Such a series of night shifts upsets the internal clock, while a single night shift would not disturb the entrained cyclicity severely.

The other solution, both theoretically sound and supported by experience, is to entrain new diurnal rhythms. It takes regular and strong zeitgebers to "overpower" the regular signals, such as light and darkness. For shiftwork this means that the same setup (such as working the night shift) should be maintained for long periods of time (several weeks, even months) and not be interrupted by different arrangements (in theory, not even by "free" days, such as on weekends). It appears that certain people are more willing and able to conform to such regular "non-day" shift regimens than others.

Health complaints of shiftworkers are often voiced and suspected, but actual negative effects are difficult to prove. Night shiftworkers have, on the average, about half an hour shorter sleep time than persons who are permanently on day shift. (However, Carvalhais, Tepas, and Mahan (1988) found that persons who permanently work the evening shift sleep about half an hour longer than persons on the day shift.) Also, persons on night shift often complain about the reduced quality of sleep that they receive during the day, with noise often mentioned as particularly disturbing.

Some authors have found statistically significant health complaints as a function of shiftwork, particularly digestive disorders and gastro-intestinal complaints, while other researchers have failed to prove significance (Cooper and Smith 1985; Folkhard and Monk 1985). Altogether, no differences have been found in the mortality of night shiftworkers as compared to workers in other shifts. However, it appears fairly clear that persons who suffer from health disturbances are more negatively affected by night shifts than by other shift arrangements. It also appears that older workers, possibly due to deteriorating health and difficulties to get sufficient restful sleep (both phenomena apparently increasing with age) may be more negatively affected by shiftwork than younger workers.

Regarding performance: the reduced quantity and quality of sleep experienced by night workers leads to the conclusion that many suffer from a chronic state of partial sleep deprivation. Negative effects of sleep deprivation on behavioral aspects have been well demonstrated, as discussed earlier. For some tasks, the interaction of circadian discrepancies between work demand and body state, and of sleep deprivation may result in significant detriments to night work performance, including safety (Monk, 1989). (Accident statistics are usually confounded by many variables in addition to the shift factor, such as the work task, worker age and skill, shift schedule, etc.)

During the first night shift or shifts, performance is likely to be impaired between midnight and the morning hours, with the lowest performance around 4:00 A.M. Such impairment, which may be absent or minimal for cognitive tasks, varies in level but is similar to that induced by "legal doses" of alcohol (Monk 1989). However, as the worker continues to do night shifts, the internal clock realigns itself with the new activity rhythms, and a daily routine of social interactions, sleeping and going to work are established.

Tolerance of shiftwork is different from person to person and varies over time. Three out of ten shiftworkers have been reported to leave shiftwork within the first three years due to health problems encountered. Tolerance of those remaining on shiftwork depends on personal factors (age, personality, troubles, and diseases; the ability to be flexible in sleeping habits, to overcome drowsiness), on social-environmental conditions (family composition, housing conditions, social status), and of course on the work itself (workload, shift schedules, income, other compensation). These factors interact, and their importance differs widely from person to person and changes over one's work life (Costa, Lievore, Casaletti, Gaffuri, and Folkard 1989; Rosa and Colligan 1988; Volger, Ernst, Nachreiner, and Haćnecke 1988). Evening shiftworkers suffer particularly in their social and domestic relations, while night shiftworkers are more affected by the conflicts between the requirement to work while physiologically in a resting stage and by insufficient sleep during the day. However, physiological and health effects are not abundant in shiftworkers who have been on such assignments for years, possibly because persons who cannot tolerate these conditions abandon shiftwork soon after trying it (Bohle and Tilley 1989).

Social interaction needs are individually different. For example, parents of small children usually need to be home for family interaction and are unlikely to accept unusual work assignments, while older persons who do not need to interact with their children so intensely may be more inclined to work "non-normal" hours.

The major problem associated with shiftwork is the difficulty of maintaining normal social interactions when the work schedule forces one to sleep during times in which social relations usually occur. This makes family relations difficult, as well as interaction with friends and participation in "public events" such as sports. Common, daily activities may not be easily executed, such as shopping or watching television.

It is noteworthy that, in different countries and cultures, certain events or conditions may be present or not, may be regularly scheduled at different times, and be considered of different value to individuals and the society at large. The southern siesta time is not commonly known in northern regions. Shops that are open continuously in the U.S.A. close in the late afternoon and stay closed on weekends in Europe. Family ties are much more important in some cultures than in others and may vary among individuals. Thus, statements regarding the effects of shiftwork on social interactions may be true in one case, but not pertinent to another. However, it is generally true that shiftwork and its consequences to the individual worker often interfere with social relations. Whether this has a demonstrable effect on well-being and performance depends on the individual case.

How to Select a Suitable Work System

The foregoing discussions should have made it clear that, if at all possible, the working hours should be from morning to afternoon. However, in many cases this "normal" arrangement is replaced by shiftwork, either covering the late afternoon and evening hours, or the night.

There is an apparently inevitable fall in performance during overnight work, related to the circadian rhythm. This is of particular concern with long periods of duty, and when the overnight work period follows poor sleep. Reports of reduced performance during the night are numerous, as summarized by Rogers, Spencer, Stone, and Nicholson (1989).

The argument for permanent assignment to either an evening or a night shift is well founded on the grounds that such a permanent arrangement allows the internal rhythms to become re-entrained according to this rest/work pattern. However, that reasoning is not as convincing as it might appear as most shift arrangements are not truly consistent or permanent, because the weekend interrupts the cycle. Furthermore, strong zeitgebers during the 24-hour day (such as light and dark) remain intact even for the person on regular evening or night shifts, thus hindering a complete re-entrainment of the internal functions. This leads to the opposite conclusion, also well founded in theory. Accordingly, it is better to work only occasionally outside the morning/afternoon period and to work only one such evening or night shift. In this case, most people are able to perform their unusual work without much detriment for this one work period, while they remain entrained on the usual 24-hour cycle. Of course, some individuals are able and willing to adjust fairly easily to different work patterns. Thus, unusual work patterns may be acceptable to "volunteers."

For crews of airplanes who must cross time zones in their long distance flights and catch some sleep at their destination before returning, several problems exist. The first is that the quality and length of sleep at the stopover location is often much worse than at home. The resulting tiredness is often masked or counteracted by use of caffeine, tobacco, and alcohol (Graeber 1988). The second problem is the extended time of duty, which includes preflight preparations, the flight period itself, and the wrap-up after arrival at the stopover. Negative effects are substantially larger after an eastward flight than for a westward direction; also, crew members over 50 years are more affected than their younger colleagues (Graeber 1988).

Recommendations for the "shift" arrangement for flight crews are fairly well established (Eastman Kodak 1986). In general but particularly when flying eastward, flight crews should adhere to well-planned timing. This should duplicate, as far as possible, the sleep-wake activities at home, meaning that the crew should try to go to bed at their regular home time and get up at their regular home time. Thus, they maintain their regular diurnal rhythm. Of course, their next flight duty should be during their regular time of wakefulness.

With respect to the length of a work shift, one general recommendation is that physically demanding work should not be expected over periods longer than 8 hours unless frequent rest pauses are available; but an 8-hour shift may be too long for very strenuous work. The same applies to work that is mentally very demanding, requiring complex cognitive processes or high attention. For other "everyday" work, durations of 9, 10, even 12 hours per day can be quite acceptable. Flextime arrangements often are welcomed by employees, possibly in combination with "compressed" workweeks, particularly if they allow extended free weekends.

Summary: shiftwork is often desired for organizational/economic reasons. Individual acceptance of shiftwork depends on a complicated balance of professional and personal concerns, including physiological, psychological, and social aspects. Of ten persons assigned to shiftwork, seven or eight are likely to stay on this schedule while the others drop out.

Pre-existing health problems may be exacerbated by shiftwork. Gastrointestinal/digestive problems are fairly frequent among shiftworkers. Workers on permanent night shifts also often complain about insufficient sleep and general fatigue, but cardiovascular or nervous diseases are not more prevalent among shiftworkers than in the general population.

Regarding work performance of shiftworkers, or accidents, little reliable information is available because on evening and night shifts other work is often performed than during the day. But work (other than cognitive) on the first night shift is likely to be impaired in the early morning hours when "fatigue" affects the worker like "legal doses" of alcohol.

Deciding which one of the many possible shift plans to select, criteria must be established to allow justifiable and systematic judgments. For example, one might establish the following requirements:

- Daily work duration should not be more than 8 hours.
- The number of consecutive night shifts should be as small as possible; best, only one single night shift should be interspersed in the other work shifts.
- Each night shift should be followed by at least 24 hours of free time.
- Each shift plan should contain free weekends, at least two consecutive work-free days.
- The number of free days per year should be at least as large as for the continual day worker.

Using these criteria, Knauth, Rohmert, and Rutenfranz (1979) discussed a large number of shift plans to select those that comply with the requirements. Following similar methods, one can carefully determine suitable work schedules which fit the given requirements and conditions (Eastman Kodak 1986).

For evening or night shifts, high illumination levels should be maintained at the workplace, such as 2000 lux or more. This helps to suppress production of the hormone melatonin which causes drowsiness. Furthermore, environmental stimuli should be employed to keep the worker alert and awake, such as occasional "stirring" music, provision of hot snacks and of (caffeinated) hot and cold beverages. The work should be kept interesting and demanding, since boring and routine tasks are difficult to perform efficiently and safely during the night hours.

The shift worker should use "coping strategies" for setting the biological clock, obtaining restful sleep, and maintaining satisfying social and domestic interactions. Unless the shiftworker is on a very rapidly rotating schedule, the aim is to re-set the biological clock appropriately to the shiftwork regimen. For example, sleep should be taken directly after a night shift, not in the afternoon. Sleep time should be regular and kept free from interruptions. Shiftworkers should seek to gain their family's and friends' understanding of their rest needs. Certain times of the day should be set aside specifically and regularly to be spent with family and friends.

SUMMARY AND CONCLUSIONS

Human body functions and human social behavior follow internal rhythms. Aside from the female menstrual cycle, the best known rhythms are a set of daily fluctuations, called circadian or diurnal rhythms. Examples are body temperature, heart rate, blood pressure, and hormonal excretions. Under regular living conditions, these temporal programs are well established and persistent.

The well synchronized rhythms and the associated behavior of sleep (usually during the night) and of activities (usually during the day) can be de-synchronized and put out of order if the time markers (zeitgeber) during the 24-hour day are changed and if activities are required from the human at unusual times. Resulting sleep loss and tiredness influence human performance in various ways. Mental performance, attention and alertness usually are reduced, but execution of most physical activities is not. Furthermore, concerns exist that disturbing the internal rhythm, such as by certain types of shiftwork, might have negative health effects. However, only gastro-intestinal problems are more frequent with workers on night shift than with persons on day work. Certainly, being excluded by shiftwork from participating in family and social activities is difficult for many persons.

From a review of physiological, psychological, social and performance behaviors, the following recommendations for acceptable regimes of working hours and shiftwork can be drawn:

- Job activities should follow entrained body rhythms.
- It is preferable to work during the daylight hours.
- Evening shifts are preferred to night shifts.
- If shifts are necessary, two opposing rules apply: (1) either work only one evening or night shift per cycle, then return to day work, and keep weekends free, or (2) stay permanently on the same shift (whichever it is).
- A shift duration of eight hours of daily work is usually adequate, but shorter times for highly (mentally or physically) demanding jobs may be advantageous; and longer times (such as 9, 10, or even 12 hours) may be acceptable for some types of routine work.
- Compressed workweeks often are acceptable for routine jobs, for example, 4 days with 10 hours.

REFERENCES

Basmajian, J. V. and DeLuca, C. J. (1985). *Muscles Alive* (5th Ed.). Baltimore, MD: Williams & Wilkins.

Bohle, P. and Tilley, A. J. 1989. The Impact of Night Work on Psychological Well-Being. *Ergonomics*, 34:9:1089–1099.

Carvalhais, A. B., Tepas, D. I., and Mahan, R. P. (1988). Sleep Duration in Shift Workers. *Sleep Research*, Vol. 17, p. 109.

Colquhoun, W. P. (1985). Hours of Work at Sea: Watch-keeping Schedules, Circadian Rhythms and Efficiency. *Ergonomics*, 28:4:637–653.

Colligan, M. J. and Tepas, D. I. (1986). The Stress of Hours of Work. *Journal of the American Industrial Hygiene Association*, 47:11:686–695.

Costa, G., Lievore, F., Casaletti, G., Gaffuri, E., and Folkard, S. (1989). Circadian Characteristics Influencing Inter-Individual Differences in Tolerance and Adjustment to Shiftwork. *Ergonomics*, 32:4:373–385.

Duchon, J., Wagner, J., and Keran, C. (1989). Forward Versus Backward Shift Rotation. *Proceedings of the Human Factors Society 33rd Annual Meeting* (806–810). Human Factors Society: Santa Monica, CA.

Eastman Kodak Company (Ed.) (1986). *Ergonomic Design for People at Work*, Vol. 2, New York, NY: Van Nostrand Reinhold.

Folkard, S. and Monk, T. H. (Eds.) (1985). *Hours of Work*. Chichester: Wiley.

Froeberg, J. E. (1985). Sleep Deprivation and Prolonged Working Hours. Chapter 6 in S. Folkard and T. H. Monk (Eds.), *Hours of Work* (67–76). Chichester: Wiley.

Gillberg, M. (1985). Effects of Naps on Performance. Chapter 7 in S. Folkard and T. H. Monk (Eds.), *Hours of Work* (77–86). Chichester: Wiley.

Graeber, R. C. (1988). Aircrew Fatigue and Circadian Rhythmicity. Chapter 10 in E. L. Wiener and D. C. Nagel (Eds.), *Human Factors in Aviation* (305–344). San Diego, CA: Academic Press.

Horne, J. A. (1985). Sleep loss: Underlying Mechanisms and Tiredness. Chapter 5 in S. Folkard and T. H. Monk (Eds.), *Hours of Work* (53–65). Chichester: Wiley.

Horne, J. (1988). *Why We Sleep — The Functions of Sleep in Humans and Other Mammals.* Oxford: Oxford University Press.

Hunter, K. I., and Shane, R. H. (1979). Time of Death and Biorhythmic Cycles. *Perceptual and Motor Skills*, 48:1, 220.

Hurrell, J. J. and Colligan, M. J. (1985). Alternative Work Schedules: Flextime and the Compressed Workweek. Chapter 8 in Cooper, C. L. and Smith, M. J. (Eds.), *Job Stress and Blue Collar Work* (131–144). New York, NY: Wiley.

Johnson, L. C., Tepas, D. I., Colquhoun, W. P., and Colligan, M. J. (eds.) (1981). *Biological Rhythms, Sleep and Shift Work.* New York, NY: Spectrum.

Kerkhof, G. (1985). Individual Differences and Circadian Rhythms. Chapter 3 in S. Folkard and T. H. Monk (Eds.), *Hours of Work* (29–35). Chichester: Wiley.

Knauth, P., Rohmert, W., and Rutenfranz, J. (1979). Systematic Selection of Shift Plans for Continuous Production with the Aid of Work-Physiological Criteria. *Applied Ergonomics*, 10:1:9–15.

Kogi, K. (1985). Introduction to the Problems of Shiftwork. Chapter 14 in S. Folkard and T. H. Monk (Eds.), *Hours of Work* (115–184). Chichester: Wiley.

Lateck, J. C. and Foster, L. W. (1985). Implementation of Compressed Work Schedules: Participation and Job Redesign as Critical Factors for Employee Acceptance. *Personnel Psychology*, Vol. 38:75–92.

Lavie, P. (1985). Ultradian Cycles in Wakefulness. Chapter 9 in S. Folkard and T. H. Monk (Eds.), *Hours of Work* (97–106). Chichester: Wiley.

Monk, T. H. (1989). Shiftworker Safety: Issues and Solutions. In A. Mital (Ed.), *Advances in Industrial Ergonomics and Safety I* (887–893). Philadelphia, PA: Taylor and Francis.

Monk, T. H. and Tepas, D. I. (1985). Shiftwork. Chapter 5 in Cooper, and Smith, M. J., (Eds.) *Job Stress and Blue Collar Work* (65–84). New York: Wiley.

Minors, D. S. and Waterhouse, J. M. (1981). *Circadian Rhythms and the Human.* Bristol: Wright.

Patkai, P. (1985). The Menstrual Cycle. Chapter 8 in S. Folkard and T. H. Monk (Eds.), *Hours of Work* (87–96). Chichester: Wiley.

Persinger, M. A., Cooke, W. J., and Janes, J. T. (1978). No Evidence for Relationship Between Biorhythms and Industrial Accidents. *Perceptual and Motor Skills*, 46:2:423–426.

Rechtschaffen A. and Kales, A. (1968). *A Manual of Standardized Terminology, Techniques, and Scoring System of Sleep Stages in Human Subjects.* UCLA Brain Information Services, Los Angeles, CA.

Rogers, A. S., Spencer, M. B., Stone, B. M., and Nicholson, A. N. (1989). The influence of a 1h nap on performance overnight. *Ergonomics*, 32:10:1193–1205.

Rosa R. R. and Colligan M. J. (1988). Long Workdays Versus Rest Days: Assessing Fatigue and Alertness with a Portable Performance Battery. *Human Factors*, 30:3:305–317.

Tepas, D. I. (1985). Flextime, Compressed Workweeks and Other Alternative Work Schedules. Chapter 13 in S. Folkard and T. H. Monk (Eds.), *Hours of Work* (147–164). Chichester: Wiley.

Tepas, D. I. and Monk, T. H. (1986). Work Schedules. Chapter 7.3 in Salvendy, G. (Ed.). *Handbook of Human Factors* (819–843). New York, NY: Wiley Interscience.

Volger, A., Ernst, G., Nachreiner, F., Haenecke, K. (1988). Common Free Time of Family Members in Different Shift Systems. *Applied Ergonomics*, 19:3:213–128.

Wever, R. A. (1985). Men in Temporal Isolation: Basic Principles of the Circadian System. Chapter 2 in S. Folkard and T. H. Monk (Eds.), *Hours of Work* (15–28). Chichester: Wiley.

FURTHER READING

Folkard, S. and Monk, T. H. (Eds.) (1985). *Hours of Work*. Chichester: Wiley.

Hunter, K. I., and Shane, R. H. (1979). Time of Death and Biorhythmic Cycles. *Perceptual and Motor Skills*, 48:1:220.

Johnson, L. C., Tepas, D. I., Colquhoun, W. P., and Colligan, M. J. (eds.) (1981). *Biological Rhythms, Sleep and Shift Work*. New York, NY: Spectrum.

Horne, J. (1988). *Why We Sleep — The Functions of Sleep in Humans and Other Mammals*. Oxford: Oxford University Press.

Minors, D. S. and Waterhouse, J. M. (1981). *Circadian Rhythms and the Human*. Bristol: Wright.

Rechtschaffen A. and Kales, A. (1968). *A Manual of Standardized Terminology, Techniques, and Scoring System of Sleep Stages in Human Subjects*. UCLA Brain Information Services, Los Angeles, CA.

INDEX

A

absorption, 124, 129, 160
absorption coefficient, 155
acceleration, 64, 83
acclimation, 166, 167
acclimatization, 158, 163, 166, 167, 169
actin, 53-56
action potential, 55, 57, 58, 77, 112
activity, 178
adaptation, 75, 168
adenosine diphosphate, ADP, 132
adenosine triphosphate, ATP, 131
adipose tissue, 129, 134, 162
afferent, 75
afferent pathway, 73
aerobic, 56, 132
aerobic capacity, 146
aerobic metabolism, 130
age, 24-26, 28, 38, 41, 106, 110, 139,
    168, 180, 185, 194, 195
aging, 24
agonist, 58
air, 104, 171
air collecting system, 149
air flow, 105
air movement, 158, 160, 167
air temperature, 160
air humidity, 160
air layer, 156
air velocity, 169
airway, 106
Alfonso Borelli, 82
all-or-none, 58, 77
alveoli, 104
ambient temperature, 160
amplifier, 66
anabolism, 129
anaerobic metabolism, 56, 58 130, 132,
    139, 141
anatomical landmark, 3-5
anatomical position, 3
ancestry, 24, 25
antagonistic muscle, 56, 58, 59 162
antechamber, 112

anthropology, 2
anthropometer, 3, 7, 8
anthropometric data, 2, 11, 14, 18, 21
anthropometric data bank, 11
anthropometric information on the U.S.
civilian population, 18
anthropometric survey, 26
anthropometric technique, 8
anthropometric term, 34
anthropometry, 1-3
aorta, 112
apathy, 162
Archimedes Principle, 87
arteriole, 54, 113
artery, 54, 161
articulation, 8, 37, 40, 41, 82, 86
assimilation, 124, 129
asthenic, 10
athletic, 10
ATP-ADP metabolism, 53
atrium, 112
audition, 75
autonomic nervous system, 74, 112
average, 12
average person, 13, 28
average skin temperature, 159
axon, 55, 57, 76, 77

B

back injuries, 45
back pain, 45
backward rotation, 180
basal ganglia, 75, 78
basal metabolism, 139, 143, 146, 164
baseline, 134
basic Rest-Activity Cycle, 182
bicycle ergometer, 135, 136, 139
bio-rhythm, 177
biomechanics, 2, 81-83
biomechanical model, 82
birth rate, 28
blood, 103, 104, 109-112, 114, 116, 129,
    156